SALT LAKES

SALT LAKES

An Unnatural History

CAROLINE TRACEY

W. W. NORTON & COMPANY

Independent Publishers Since 1923

For Mariana and Lázaro

And for all creatures who love salt lakes

It was as though each plainsman chose to appear as a solitary inhabitant of a region that only he could explain . . .

—Gerald Murnane, The Plains

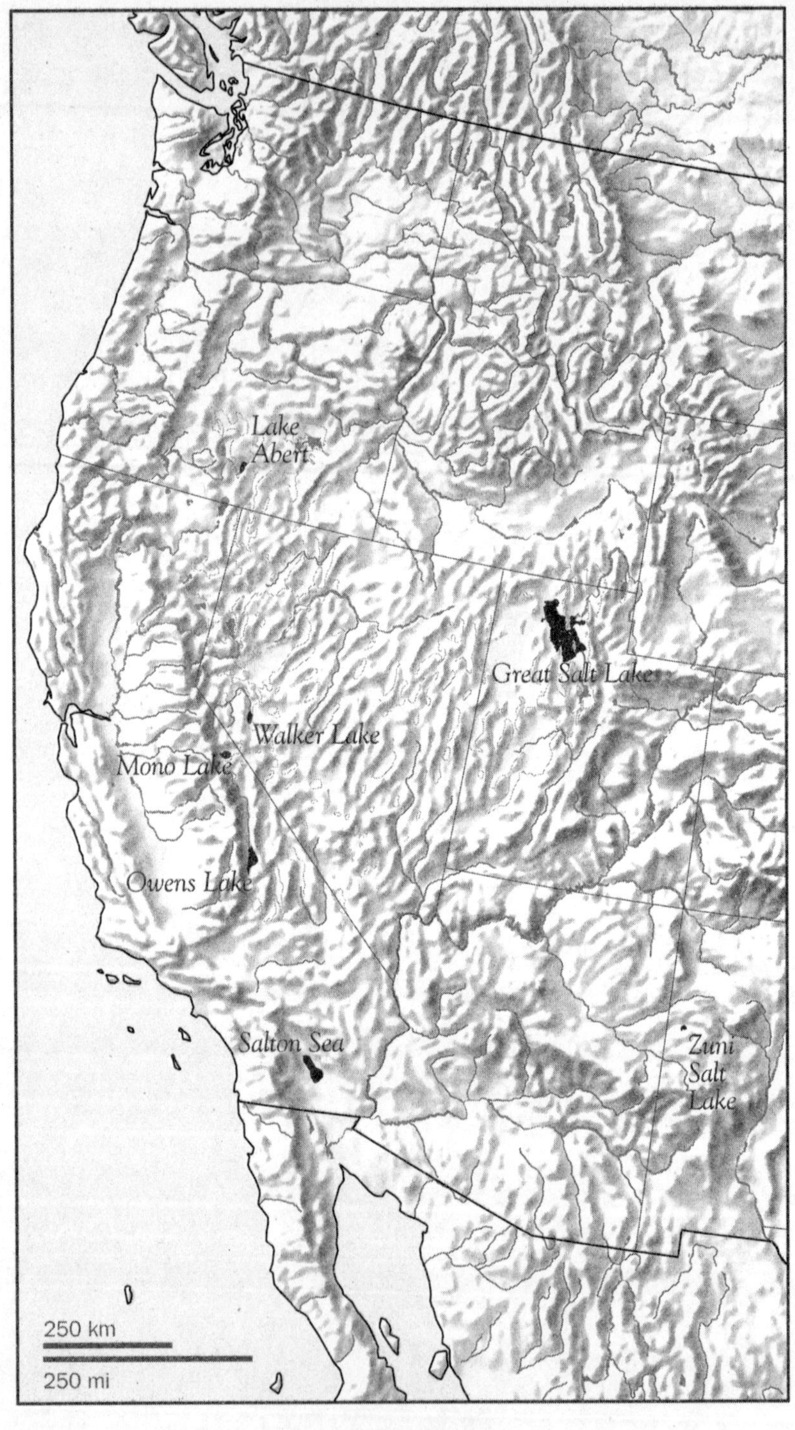

Lake
Abert
Great Salt Lake
Walker Lake
Mono Lake
Owens Lake
Salton Sea
Zuni
Salt
Lake
250 km
250 mi

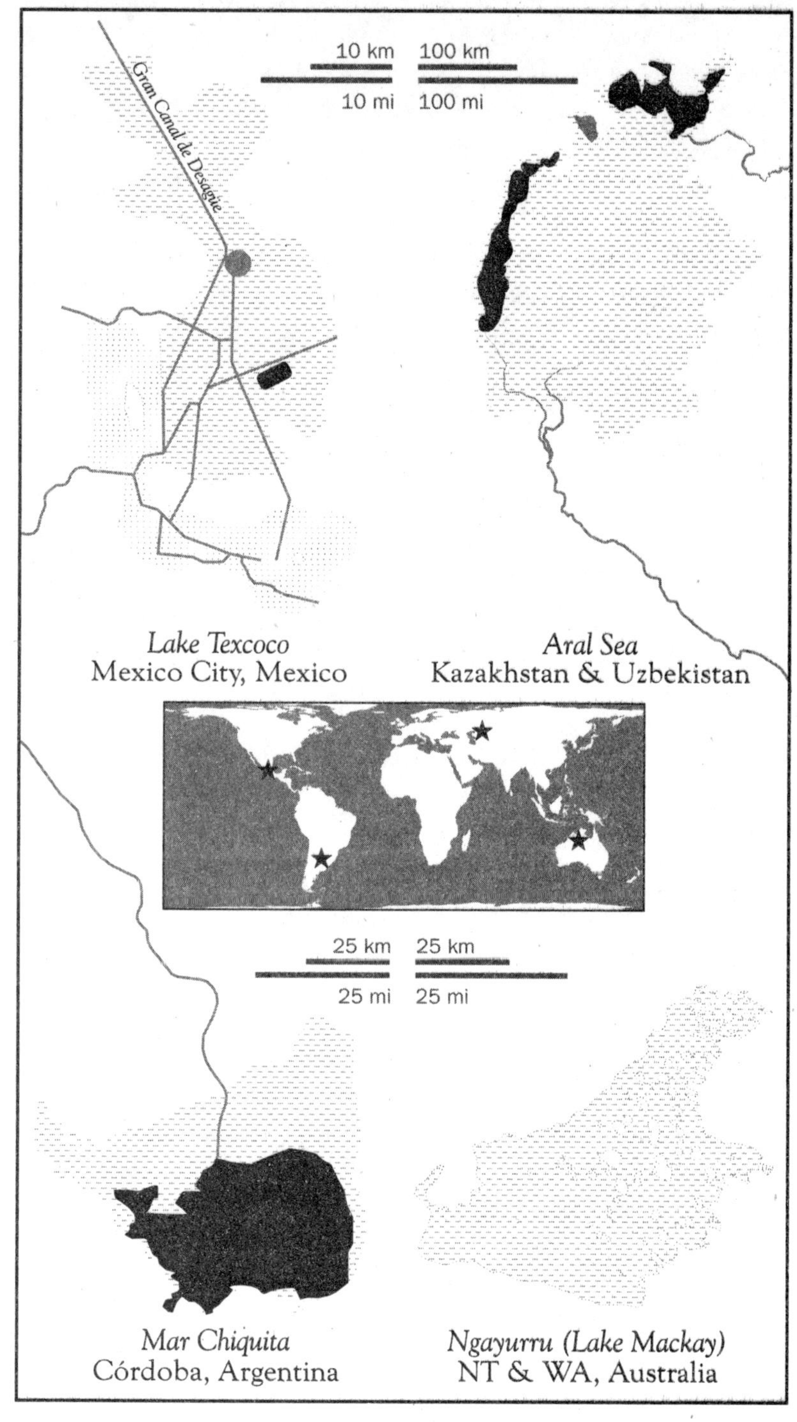

Lake Texcoco
Mexico City, Mexico

Aral Sea
Kazakhstan & Uzbekistan

Mar Chiquita
Córdoba, Argentina

Ngayurru (Lake Mackay)
NT & WA, Australia

CONTENTS

Prologue
1

I. DIVERSIONS

CHAPTER 1
Closed Basins and Sacred Lakes
Great Salt Lake, Utah, USA
11

CHAPTER 2
Richness in Damaged Places
Salton Sea, California, USA
25

CHAPTER 3
Imperfect Recovery
Aral Sea, Kyzylorda, Kazakhstan
45

II. RECALIBRATIONS

CHAPTER 4
The Common Good
Mono Lake, California, USA
65

CHAPTER 5

Restitution
Zuni Salt Lake, New Mexico, USA
87

CHAPTER 6

A River Passes By Here
Lake Texcoco, Mexico City, Mexico
119

III. SALINE FUTURES

CHAPTER 7

The Ecosystem that Lawsuits Made
Owens Lake, California, USA
151

CHAPTER 8

Willing Sellers and the Public Trust
Walker Lake, Nevada, USA
173

CHAPTER 9

An Act of Attention
Lake Abert, Oregon, USA, and Mar Chiquita, Córdoba, Argentina
189

CODA

The Ephemeral Forever
Lake Mackay, Western Australia/Northern Territory, Australia
211

Acknowledgments 225

Notes on Sources 228

Bibliography 244

                                        CONTENTS

SALT LAKES

PROLOGUE

WHEN I WAS TWENTY-FOUR, I WORKED AS A COWBOY in New Mexico. After rainstorms, the brown-red earth turned white in some places, especially at the edges of the creek that ran through the ranch. I didn't understand where it came from. I only knew that as a queer woman searching for home in a world fast being remade by climate change, I kept finding my way to salt.

The earth had fascinated and worried me since childhood. Since I could remember, the world had opened itself up to me in my ravenous reading. Yet at the same time, I saw its destruction everywhere I looked. The gray rooftops of Denver's suburbs stretched further and further onto the rich, dark soils of the prairie; less and less of the snow on the mountains lasted through the summer with each passing year. On the city's beautiful sunny days, my mom would repeat the old *Denver Post* slogan, "'Tis a privilege to live in Colorado." Yet in the summer after the sun went down, she would note her surprise at no longer having to put on a sweater to sit outside.

We never questioned that we belonged where we lived, or that we were rooted there: our ancestors had been some of the state's earliest white settlers, had helped build its irrigation ditches and political institutions. Yet to me, it seemed that we were squandering the privilege. I wanted to live in a society that respected its surroundings. To be able to imagine such a way of life, I thought, I needed to find words that could articulate the sublime meaning that the vast, arid region around us seemed to possess, and that in turn could develop a land ethic to teach us how to live there.

The August I was fourteen, I bought a copy of Wallace Stegner's *The American West as Living Space*. I liked the cover's grayscale image of a man standing on a gargantuan irrigation canal, the kind of infrastructure that, as a Westerner, made me feel at home, even as I recognized it as symbolic of the trouble we were in. No other book I had read articulated my feelings about the region so well—both my sense that the place where I was growing up was exceptional and my fear at the rapidity with which it was changing. "What makes that western space and distance?" Stegner asked. Aridity, he answered, again and again. The condition of dryness was what gave the West its distinctive character and its distinctive challenges alike.

Reading Stegner convinced me that I, too, could capture the importance of the West in writing. I scoured used bookstores for his work, collecting as many of his many books as I could find. I revered the political influence he had held in the region, and grandiosely imagined that if I worked hard enough, my words could help solve the environmental problems of my own time.

The spring before I graduated from high school, a new biography of Stegner, written by Western environmental historian Philip Fradkin, was published. Fradkin quoted a letter that Kentucky writer Wendell Berry sent to Stegner after the publication of his 1971 novel *Angle of Repose*, about a family following mining, hydrology, and construction jobs across the West and Mexico. Berry told Stegner that his book was "more Russian than American; it made me think of Tolstoy and Pasternak."

Perhaps, I thought, I could hew the vocabulary I needed for the American West out of Russian tools. So I majored in Russian, a literature that indeed opened up my favorite kinds of landscapes. In Platonov's short stories, I found decaying boomtowns like the southeastern Colorado community where my grandfather had grown up, a once-prosperous sugar beet center that had sold its senior water rights to a Denver suburb. In Saltykov-Shchedrin's *The Golovlyov Family*, with its description of staring vacantly "at the seemingly boundless fields, stretching away into the remote distance," I found the hazy, inert bleakness I came to know while visiting my college boyfriend's family in California's Central Valley. In Chekhov's personifications of storms breaking over the steppe, his evocations of wind catching tumbleweeds and dust spiraling, in his ambivalent equation of planting trees with virtue, I found both the ethic of moral responsibility to place I was searching for and the intensity of experience I knew from being in rich landscapes.

What felt most important was the way the writers imbued the geographies around them with spiritual depth. They made places not just superficially beautiful, as American nature writing sometimes seemed to do, but sites where philosophy could be found. They saw processes like building power lines and irrigation canals not only as mundane obligations in the development of a society, but as moments on which to hammer out an understanding of the type of people who would populate it—their orientation to the universe, the contents of their souls.

Then, I started crossing paths with salt lakes. In the months after graduating from college, my boyfriend, Dylan, and I began driving around the Great Basin. The Great Basin is a 220,000-square-mile area, spanning six states of the American West, that has no outflow to the ocean. It's not just one big closed basin, but a massive complex of more than a hundred. On the map, it looks empty, brown, and dry. The area is "well known only to a few gamblers, professional criminals, movie stars, divorcees, and, of course, the people who live there," poet Ed Dorn writes in his 1966 book *The Shoshoneans*.

After my Colorado childhood, I believed I knew the West. But the West Dylan knew was a different one. He had stories about leading mule trips for tourists on the backside of Yosemite, about Nevadans asking him whether he was "packing heat" when he went out for a run, about hot springs and sand dunes. We went together to the Salton Sea, to Owens Lake and Deep Springs Lake. We zigzagged through the innumerable valleys of Nevada, practically every one of which has its own salt lake or at least the crust left behind by a former one.

The lakes' beauty captivated me, and their strangeness piqued my curiosity. They appeared suddenly in places I never expected to find huge pools. From afar, their palette of glistening blue water, white salt flats, green wetland edges, and fuchsia and emerald microbial life turned the horizon into a painting. The way the salts heavied the chemistry of the water made them reflect the surrounding landscape like a mirror.

Salt lakes form when water collects at the lowest point of closed basins. Scientists call these bowls in the landscape endorheic basins, from the ancient Greek words for "internal" and "flow." They mainly exist in desert regions of the world; in places with more water, the currents gradually carve a way out. In contrast to ordinary valleys with an outlet, and in contrast to freshwater lakes—which are, in the words of John McPhee, an "aneurysm in a river"—in endorheic basins, water gathers but has nowhere to flow. It can only evaporate away.

When it does, it leaves behind the salts that were suspended in it. The salts aren't just sodium chloride, or table salt, but a variety of dissolved minerals that depends on the geology of the watershed. Their concentrations vary, too: some endorheic lakes are just a tenth as saline as seawater; others are ten times saltier.

Most permanent salt lakes—perennial lakes, in the language of scientists—were once much larger bodies of water. During the Pleistocene era, when earth's climate was cooler and wetter than it is now, large lakes covered what are now arid regions. When the climate warmed about twelve thousand years ago, they evaporated into the smaller bodies we know today, leaving the salts behind, concentrated and suspended in the remaining water. Great Salt Lake is a remnant of Lake Bonneville, whose fringes extended all the way into Idaho and Nevada; its erstwhile shorelines are visible in the Salt Lake Valley, like shelves protruding from the mountains. Lake Lahontan spanned much of what is now Nevada, leaving behind Pyramid Lake, Walker Lake, and the now-dry Lake Winnemucca when it evaporated.

At smaller, temporary lakes, all that's left of the ancient bodies that preceded them is a playa—a salt crust surrounded by a mudflat—where water sometimes pools. In temperate and high-altitude regions, these playa lakes are seasonal: they fill in the spring, when the snow on the surrounding mountains melts and the rivers swell and rush downhill, and evaporate over the course of the summer. In hotter, lower-elevation regions and those closer to the equator, these temporary lakes are categorized as ephemeral, filling only after big rainstorms, and leaving behind dry expanses the rest of the time.

Since most of the salt lakes in the United States are located in the Great Basin, many are found on ancestral homelands of Paiute tribes. The Paiute understood the lakes' hydrology. "Our ancestors always maintained, in our legends and our stories, that the lakes are connected," Wilson Wewa, the Warm Springs Paiute Tribe's oral historian, told me. "They used to tell us stories about creatures that used to live in the lakes. White people would make expeditions to try to put to rest our 'superstitions' and our old people used to say, 'They aren't going to find them because they went underground to a different lake.' Knowing what I do today, about how Lake Lahontan covered almost the whole state of Nevada, and about how the lakes in the Great Basin are connected by the aquifer, I think to myself, our past people knew more about this connection than people give them credit for."

Being endorheic makes salt lakes much more sensitive to change than freshwater lakes. They fluctuate naturally in response to the different amounts of snow and rain that fall each year, but rely on a balance of inflows and evaporation that stays within a certain range. In recent decades, humans have diverted so much water from the rivers that feed the lakes—mainly in order to irrigate fields of alfalfa, cotton, and other commodity crops—that we've upset that balance. More water evaporates from the lakes than enters them. At the same time, global warming from the combustion of fossil fuels is decreasing snowpack, changing weather patterns, and making evaporation work faster. There's less water in the system, less of it reaches the lakes, and once it gets there, it leaves more quickly.

As a result, around the world, salt lakes' water levels are on a steep downward trend. Great Salt Lake, the Aral Sea, the Caspian Sea, Lake Texcoco, Lake Urmia, Lake Rusanda, Lake Cuitzeo, and dozens of others are all drying up. Some have disappeared altogether. That's a problem because the lakes are the endpoint of larger water cycles and harbor important food chains. When the strange, hidden salt lakes start dying, it means entire ecosystems are in bad shape.

Far more of us than we realize are implicated in this disappearance, and the consequences will be far-reaching. If you live in Los Angeles, your drinking water comes at the expense of two salt lakes, Owens and Mono. If you live in San Diego, your drinking water once fed the Salton Sea, the residents of whose surrounding area are now plagued by a novel form of asthma due to dust from its drying lakebed. If you live in Mexico City, your life takes place atop the bed of a drained salt lake. If you drive an electric car or use a smartphone, it is powered by a battery whose lithium was likely mined from a salt lake high in the Andes. Perhaps most shockingly, all the water that is diverted from salt lakes and evaporates off of farmland in endorheic basins ends up falling as precipitation in other, wetter parts of the world, and scientists have calculated that it adds up to the second-largest cause of sea level rise, after melting glaciers.

When I started researching salt lakes, I didn't yet know that arid landscapes are full of salts. I didn't know that those salts flow through capillaries in the soil to bloom after water seeps in. I hadn't driven around the Great Basin enough to understand that there's a shimmering salt crust at the base of every one of its valleys. Each new encounter with salt felt like an errant, miraculous surprise.

But I kept finding them—or they kept finding me. We were witnessing one another. As I started to document scientists' and activists' efforts to bring the lakes to a stable state, the lakes, in turn, began to watch me grow up, to figure out what kind of life I wanted to lead and what kind of meaning I would make of it.

By now, I've been traveling to and documenting these lakes for more than a decade, building relationships with the scientists and activists who are working together to keep the lakes' ecosystems intact and to keep dust down. I've learned in the process that the decline of salt lakes is a solvable one: unlike other types of ecological restoration, which involve engineering and complex chemistry, or solving climate change, which requires changing nearly everything about contemporary life, saving salt lakes simply requires adding water. Still, the coming years are a political and environmental mystery: Will we do what is necessary to save the lakes?

The question of whether salt lakes will be saved is also the question of whether we'll find a land ethic that allows us to continue living in the West and other arid regions. Even if saving the lakes just requires adding water, doing so will require reimagining water law, agricultural practices, and societal priorities. Having the imagination to enact those changes requires enriching our sense of the world and loving life in all its varied, queer forms, small and large.

I think those changes are possible. This book is the story of how the salt lakes taught me that optimism. It's about self-discovery, finding joy in the quotidian, and opening our eyes to the complex, hidden vitality that flourishes in overlooked places, from remote lakes to our everyday surroundings.

I.
DIVERSIONS

CLOSED BASINS AND SACRED LAKES

Great Salt Lake, Utah, USA

IN LATE JANUARY 2024, I GATHERED WITH 1,200 OTHER people at the steps of the Utah Capitol to rally in support of Great Salt Lake, the largest salt lake in the Western Hemisphere. Just over a year earlier, it had receded to a record low size of 941 square miles, little more than half its historic average. Too much water from the rivers that fed the lake was being diverted to irrigate fields. The bays where the rivers once mouthed open into wide, reedy deltas were now dry. The salts were becoming too concentrated even for highly adapted creatures to survive. Without significant political action, scientists warned, the lake could dry up in just a handful of years.

Timed to coincide with the start of the state's legislative session, the scene was colorful and jubilant. A group of artists had made props that seemed to belong on a theater set: full-body brine shrimp getups made of pink and orange felt, papier-mâché hats that turned their wearers' heads into those of shorebirds, painted cardboard cutouts of hawks that stuck out of the crowd on tall wooden dowels, canvas banners with phrases like "Defend our Future," and long strips of blue cloth painted to look like waves.

Those gathered included young environmental activists from Salt Lake City's polluted west side, queers with "Support Trans

Healthcare" and "Free Palestine" patches sewn onto their jean jackets, skiers wearing bright fleece pullovers, athletes in running leggings, and gray-haired men with kerchiefs reading "LDS Earth Stewardship" around their necks. I stood alone at the back, wearing a hunter green jacket, a plaid wool scarf, and Adidas Sambas with a hole in the big toe. I wished I'd worn warmer socks.

The speakers alternated between peppy youth activists who led chants and songs and contemplative older speakers. Three Indigenous leaders—Darren Parry of the Northwestern Band of the Shoshone Nation, Forrest Cuch of the Ute Nation, and Beverly Harry of the Navajo Nation—shared the lake's importance to them with the crowd. Cuch, the former executive director of the Utah Department of Indian Affairs, said that once, the lake was "our sacred lake"—that of Utah's Indigenous peoples. Now, he continued, gesturing to everyone gathered, it was *our* sacred lake—a shared responsibility and place of meaning. He led the group in a chant: "Our sacred lake. Our sacred lake!"

Writer Terry Tempest Williams took the podium wearing a hat resembling an eared grebe, a bird whose population depends on Great Salt Lake. Williams's 1991 book *Refuge: An Unnatural History of Family and Place* chronicles the lake's 1987 record high water year, which coincided with her mother's death from breast cancer, and reflects on the ways that generations of her Mormon pioneer family are intertwined with Great Salt Lake and the Great Basin around it. "Birds are prayers with wings," she told the crowd, gesturing to her hat.

Finally, the event closed with ecosystem scientist and Brigham Young University professor Ben Abbott. In keeping with the exuberance of the rally, he was wearing red jeans and a red cable-knit sweater. His twelve-year-old son had stood with him on the capitol steps for the previous hour and a half.

I'd met Abbott over Zoom the year before, when I wrote an

article about environmentalists from the Church of Jesus Christ of Latter-day Saints who hoped to persuade their church to take action regarding the Great Salt Lake. As an institution, the Mormon church is notoriously conservative. Many people see its theology as anti-environmental, too, because of its focus on preparing to spend eternity in heaven. But Abbott was part of a contingent that believed its scripture to be supportive of environmental causes. During our call, he had spoken eloquently about the ways that he saw activism and science as compatible with each other and with faith, even when others in the church didn't.

Before the 1970s, I learned, many church leaders believed that conservationist practices were important ways of living out their faith. Many were anti-hunting; one church president was a vegetarian, and another stated that his favorite song was a children's ditty called "Don't Shoot the Little Birds." The Doctrine and Covenants, one of Mormonism's holy texts, includes the "law of consecration," which forbids using more than one needs, including of natural resources. Meanwhile, the Word of Wisdom, the text that famously prohibits Mormons from consuming alcohol and coffee, also dictates that meat be consumed sparingly, and fruits and vegetables eaten in season. "The drying of the Great Salt Lake is being driven primarily by growing alfalfa for feed for animals," Abbott had said. "If we were to follow that clear guidance in scripture to have a plant-based diet, we wouldn't be in this situation." Later, he told me he imagined creating a team of "water missionaries" to knock on doors across Utah, teaching people about the importance of conservation. He had a broad, intense smile, and unflagging optimism. And he had fully endeared himself to me shortly after the interview, when he posted to Twitter a video of himself playing Sufjan Stevens's "To Be Alone With You" on the guitar.

Since then, it seemed that all of Salt Lake City had fallen under his spell. At an event the previous night, Williams had declared that she wished he would be the state's next governor. "It's all hands on deck," she had said of the Great Salt Lake fight, a phrase I recognized as one of Abbott's standards.

Now, Abbott opened by asking who had attended the rally for the lake the previous January. "It's only been a year, and yet I feel a different energy," he said. "What I feel here is something deeper than hope. For me, I call it faith. Faith is more than belief—it's the ability to face dread and let it be overcome by confidence; the ability to mobilize action, to tear down dams and let water flow out of them."

I'd also seen a change in the previous year. For decades, Utahns had scorned the Great Salt Lake. "People saw it as *worse* than the desert," *Salt Lake Tribune* reporter Leia Larsen told me over lunch later that month. "It was this pointless, smelly lake whose water you can't even use."

Larsen was one of the first reporters to draw attention to the fact that the lake was shrinking, after operators of dinner cruises on the lake told her the retreating water was affecting their business. For years, its decline remained a local story, and one about which few locals were even concerned. But when the water began to career to record low levels, artists, students, physicians, outdoor enthusiasts, and others all began to mobilize. In fall 2022, a group of youth activists held a funeral and die-in for the lake, consisting of a silent procession in which they carried gravestones hallowing the dying lake and a service where they read poems and eulogies honoring it. Months later, an elementary school classroom successfully fought for the brine shrimp to be named Utah's State Crustacean, a symbolic act important because, as one of the sixth-graders told me, if the lake dried up, "we'd have a city named after a pile of dirt." Microbiologist Bonnie Baxter filmed herself reading an obituary for the lake on its shore. The lake had become an urgent, beloved matter of collective struggle.

As the lake dried, its ecosystem was collapsing. The losses would be consequential. Salinity makes Great Salt Lake uninhabitable for all but highly adapted species. Algae and bacteria and archaea color portions of the lake green and reddish-purple. Brine shrimp, which I first encountered as sixth-grade class pets in their novelty form—sea monkeys—float in the water column. Together with brine flies, the shrimp feed the twelve million migratory birds that seek safe harbor in the lake's wetlands, including black-necked stilts, snowy plovers, western sandpipers, white-faced ibis, and peregrine falcons.

Some of those species, like Wilson's phalaropes and American avocets, evolved to live at saline lakes, learning to feed in ways that avoid swallowing salt and developing a specialized anatomy to deal with the minerals they do ingest. The birds are attracted to the salt lakes precisely because they're so inhospitable to most creatures— because they are, as ornithologist Margaret Rubega put it to me with fondness, "freakish habitats in the landscape."

Brine shrimp and brine flies are the only invertebrates at most salt lakes, and "not many birds want to make a living on shrimp," as another scientist told me. But for the few who do, the invertebrates are abundant and easy to eat. The buffet of shrimp and flies and the relative absence of predators makes the lakes a safe place for the birds to rest, molt their feathers, and double their body weight in order to be able to fly south for the winter—all the way to Argentina, in the phalaropes' case.

Now, the increased salinity was making it harder for microbes to survive, which in turn made it harder for tiny brine shrimp and brine flies to reproduce, which meant there wasn't enough food to feed the birds. The carcasses of eared grebes—the bird on Terry Tempest Williams's hat, 90 percent of whose entire world population stops at Great Salt Lake every spring—piled up dead on the shore, looking like deflated toys. Even for we humans, at the tippy-top of the food chain, the effects threatened to be severe. The increasingly dry lakebed released more and more dust in storms that, scientists warned, could eventually make Salt Lake City's air lethal to breathe.

BECAUSE OF THE PRESENCE of the Church of Jesus Christ of Latter-day Saints, Salt Lake City has an aura unlike that of any other place in the United States. The US's major homegrown religion, the church was founded in 1830 in upstate New York after Joseph Smith found a set of plates engraved with the Book of Mormon—an account of Israelites who migrated to the Americas in ancient times.

For Mormons, Salt Lake City is Zion: the church's spiritual and political center, a gathering place for the faithful. It's a place where time and space are ordered by faith; church, politics, and the rhythms of daily life are synchronized. The city, along with practically every other town in the state, is laid out on a coordinate grid at whose central point, 000,000, is the temple—the "sanctuary that connected earth and heaven," in the words of historian Benjamin Park. Utah's state routes are marked with the outline of a beehive, a reference to a Mormon symbol that represents the cooperation and hard work necessary to build the Kingdom of God. Religious words like "sealed" and "stewards" appear in quotidian, civilian conversation with a frequency jarring to outsiders. Wallace Stegner wrote of the city: "To the devout it is more than a place; it is a way of life, a corner of the materially realizable heaven; its soil is held together by the roots of the family and the cornerstones of the temple. In this sense, Salt Lake City is forever foreign to me, as to any non-Mormon."

The story of how Great Salt Lake got into such trouble, then, is in large part a Mormon story. But it is also a template for the story of the decline of other salt lakes. That's because Mormon society can be seen as a microcosm of American society at large, especially society in the American West. The story of how Mormon settlement and violence and belief transformed into contemporary infrastructure is also the story of water in the West.

Mormons first came to the Salt Lake Valley in 1847, led by Brigham Young, the church's second president. They chose to settle in the Salt Lake Valley, in part, because it was an oasis. Abundant creeks rolled down from the mountains to Great Salt Lake and the freshwater Utah Lake upstream. The soils were rich and dark. The grass grew waist-high. It seemed clear that the valley had been prepared for them by God.

Great Salt Lake was part of this sacred vision. The fact that the valley had one freshwater lake and one saltwater lake—Utah Lake and Great Salt Lake—reminded the settlers of Israel, with the Sea of Galilee and the Dead Sea. They named the river that connects them the Jordan River, anchoring their community in holy time and space.

They took a dip in the lake within days of arriving, and it astonished them. "We cannot sink in this water. We roll and float on the surface like a dry log. I think the Salt Lake is one of the wonders of the world," wrote Orson Pratt, a member of the first pioneer company. It became part of daily life. Women made pickles in vats of briny water harvested from the lake, and discovered that the salt efflorescences that bloomed on the ground, which they called salaratum, contained the right mix of sodium bicarbonate to make soda bread.

The following year, the salt lake saved the settlers. A plague of insects was tormenting their crops, threatening the community's food supply. They tried burning their fields, even though it meant losing any chance of a harvest. The bugs—a species of katydid now known as Mormon crickets—kept coming. The settlers fasted and prayed for three days. Then came the miracle. Hordes of seagulls flew in from Great Salt Lake and devoured the crickets, salvaging the settler's crops.

The church continued to bring more settlers to the region, not only from the Midwest and the East Coast but also from Europe, especially England and Ireland. Dylan and I discovered after meeting each other that we both had forebears who had walked from Iowa to Salt Lake City in Mormon handcart companies. My ancestor, named Twiss, came from Dublin via Liverpool in 1856 with his wife and three children. After my grandfather died, my uncle emailed me a scan of Twiss's diary from the five-month trip, full of fainting and strange British timekeeping. "21st June: Started at 7 1/2. Camped at South Skunk Creek," he wrote. "Travelled 14 miles. A child died this morning and was buried under a tree."

When my ancestors finally arrived, however, Salt Lake City disappointed them. They left the church within a year and traveled back the way they had come, settling in Florence, Nebraska. Dylan's family arrived several handcart companies later, and remained in Utah and Idaho until they moved to California when his mother was ten. Though they left the church shortly thereafter, above the fireplace in his aunt's house still hangs a print of a painting of the handcart trail.

The settlers were building their sacred geography in a region inhabited by people who already had a relationship to holy time and space. Utah and Timpanode tribes lived south of the lake; Shoshones and Snakes lived to the north. The settlers encountered them, more than anywhere else, at bodies of water.

They felt prepared for these encounters. Mormon belief had a specific place for Native Americans: they were Lamanites, a "fallen branch of Israel" that, once redeemed, would "help the Latter-day Saints usher in the Last Days," in the words of historian Jared

Farmer. But the tribes were more resistant to being educated about this role than the settlers expected. When coexistence became challenging, the settlers tended toward violence. In early 1850, Brigham Young authorized the settlers' militia to carry out an assault. "They must either quit the ground or we must," he told church leaders. Later that year, he sent a letter to Washington, DC, asking the federal government to remove the tribes. "They are doing no good here to themselves or any body else," he wrote.

When the US Army arrived to bolster the settler militias in the mid-1850s, Utah became the theater of full-scale Indian War. In January 1863, the aggressions culminated in the Bear River Massacre, in which soldiers and settler militiamen killed between 250 and 400 Shoshones northeast of the lake—at that time, the bloodiest attack on Native people west of the Mississippi. By the end of 1865, the army had deported the region's remaining Native people to reservations in eastern Utah.

As the settlers severed the tribes' relationships to the sacred lake, they also dissolved their own. To focus their energies on building Zion, at church each Sunday parishioners were assigned work groups and chores: planting, plowing, building the temple and the gridded city that surrounded it. Irrigation carried particular importance in their vision for the sacred city. One of the community's first tasks was to dam the shallow, eight-foot-wide City Creek and divert its water into a hand-dug ditch that would feed newly plowed potato fields. The settlers learned hydraulic engineering by trial and error, using glass jugs of water as levels and jerry-rigging the headgates that opened and closed to let water into their ditches. Within twelve years, they were harvesting 130,000 bushels of wheat and root crops from 17,000 irrigated acres.

Hand in hand with this engineering, they developed a system of water governance. On the east coast, settlers had copied the English system of riparian rights, granting rights to rivers and streams to those who lived along them and only allowing water to be diverted if an equal amount could be returned, to protect the waterways for transportation. In the arid Great Basin, where all farmland would require irrigation, such an arrangement was impossible. Instead, church president Brigham Young declared all natural resources to be public, including water, which he apportioned based on the amount of land an individual planned to irrigate. His goal was to divide water equitably and to make the maximum use of the supply available. Users couldn't hoard water, because that would deny their neighbors the ability to water their crops; they had to put it to "beneficial use" and were forbidden to "waste" it. The oversight of the system was hierarchical and orderly, like the church itself: users were divided into "irrigation districts" that were overseen by a "watermaster."

In part, the settlers irrigated out of the necessity of feeding themselves. But they also believed they were fulfilling a prophecy from the Book of Isaiah: that "the desert shall rejoice and blossom as the rose." They believed that the earth was under a curse and that they needed to help God in the process of regenerating it. They repeated the mantra until they believed that the valley had been a desert when they found it. The sacredness of diversions and canals soon far outshone that of the salt lake. Irrigation "was more than an economic necessity; it was a form of religious worship," writes historian Leonard Arrington. "The construction of water ditches was as much a part of the Mormon religion as water baptism."

Eventually, Young's directives regarding irrigation and the governance of water formed part of the basis of water law across the West. When Utah became a state in 1896, the districts he had assigned were codified into law. His concepts of irrigation districts and watermasters migrated into secular vocabulary, and are still common throughout the region.

Around that time, Coloradoans combined Young's concept of beneficial use with "prior appropriation"—the idea, developed by California miners, that whoever claimed a given site earliest held the senior right. As the region's states adopted variations on the "Colorado doctrine," water law in the West became an alchemy of beneficial use and prior appropriation.

When Young's concept of beneficial use, developed for the closed world of a religious community, was melded into the open, individualistic world of the frontier, its communal aims transformed. Under the Colorado doctrine, prior appropriation—not church hierarchy or communal needs—determined the seniority of rights, while beneficial use determined their size. Beneficial use was "the basis, the measure, and the limit" of a right. As a basis, the act of building a ditch and using the water created the right; as a measure, the size of the right was determined by how much you used; as a limit, the obligation to use your right acted as a check on the possibility of accumulating rights to sell them later.

The rules of the doctrine also meant that if you used less than the full amount to which you were entitled, you could lose whatever you weren't using. If you'd been granted a diversion of 30 cubic feet per second but you only diverted three, for instance, other farmers could place claims on the 27 you weren't using. In Colorado, legal decisions at the end of the 1800s turned this hypothetical possibility into a reality. The cases inspired so much fear in the minds of farmers that beneficial use came to be known as "use it or lose it." Instead of using water judiciously so as not to deprive your neighbors of their share, it became an obligation to use your full water right, no matter how much of it you needed.

The term's obverse, "waste," changed, too. Waste was the water left unused in the dictum "use it or lose it." Letting some of your water flow by was waste. Soon, any water not put to human use became waste. Letting water simply be water was waste.

Great Salt Lake was the worst water-waster of all, because any water that flowed into it was ruined by becoming saline. So, gradually, the church turned on the lake it had once considered sacred. In the 1870s, an article in the church-owned *Deseret News* described the lake as "the most wonderful body of salt water in the world" before immediately going on to declare it "worthless . . . for uses of navigation and commerce," "worthless for irrigation purposes," and "worthless for any purpose." The ideologies of use and waste, and the imperative to make the desert bloom, required of Utahns that they dam and divert their rivers before they reached the Great Salt Lake. Over time, aided by federal funds, dams and diversion canals hobbled the Provo, Jordan, Weber, and Bear rivers, sending water to farms and lawns and making Utah, the second-driest state in the United States, into the second-highest per-capita consumer of water. Even today, despite the lake's decline, state leaders hope to build even more diversion projects. The proposed Bear River Project would develop dams along the river that, after winding 350 miles through Wyoming and Idaho, provides the lake with 60 percent of its inflows. Scientists estimate that the project would lower the lake's water level by two to four feet.

As the combination of "use it or lose it" and "first in time, first in right" became the ideology of the whole settler West, the rush to divert and dam was repeated in the watersheds that feed Owens Lake, Mono Lake, the Salton Sea, Lake Abert. The collapse of the lakes, then, isn't an unfortunate accident. It's the consequence of an exhausted ideology, one mutated and taken too far, one that puts utility above all else, leaving no room for the intrinsic value of water, the plants and creatures who depend on river and lakes, or the rest of natural world.

In my quest to find philosophy in irrigation ditches, and the odyssey through the world's endorheic basins that followed, meet-

ing the environmentalists of the Church of Jesus Christ of Latter-
day Saints took me by surprise. Growing up in Denver, I had never
known that just on the other side of the Rocky Mountains, there
were hundreds of thousands of people—people from whose culture I knew part of my family came—whose religious belief gave
them a way of thinking about the landscape of the West that could
transcend the earthly, the momentary, the economic. Though we
ordered the world through different belief systems, they were my
kin in the powerful conviction that this was how we ought to be
speaking about salt lakes, our arid homes, and the earth in general.
"Why, given the stunning beauty of a place like the Wasatch front,
do so many of us *not* feel this amazement?" writes Brigham Young
University English professor George Handley. "We certainly seem
to have been in a hurry to engineer ourselves out of this state of
stupefaction before the cosmic mystery . . . as if the strangeness
of finding us here on an earth of unimaginable age and of num-
berless forms somehow diminishes or even terrifies us."

As I stood on the stone esplanade in front of the Utah capitol, my coat pocket buzzed with a text message. It was Patrick
Donnelly, the Great Basin director for the Center for Biological
Diversity and one of the most tireless advocates for the region's
saline ecosystems. "We are at an event with a thousand people
chanting 'brine shrimp, salty water,'" he wrote. "This couldn't
be any more up our alleys." I smiled. It was true. For those of us
who had loved salt lakes in isolation for a long time, sharing in
the collective energy was astonishing. Our earthly future, I knew
deeply and resolutely, depended on loving the lakes in all their
mysterious, freakish, sacred beauty.

RICHNESS IN DAMAGED PLACES

Salton Sea, California, USA

W E PARKED IN THE WILDLIFE REFUGE LOT AND walked toward the bright water. Everything else in the landscape around us—the light gray mountains, the gray-green tamarisk bushes, the white ground—was baked by the harsh sun into a pastel palette. The beach crunched under my feet. I looked down, expecting to see shells. The shells looked back at me with eyes and elongated foreheads. The shore was composed of fish bones. I recoiled.

The first salt lake I visited was the Salton Sea. It was February 2014, the height of a historic drought in California. My boyfriend, Dylan, and I had graduated from college together that December, and he had taken a job in Los Angeles conducting a research project for a large art gallery. I wrote a draft of a novel, put in a few hours a day at a copywriting job for a sketchy startup, volunteered at a school garden, and drained the savings that I had hoarded since high school, waiting for a period like this, a chance to learn to be a writer.

After four years reading thick, philosophical Russian novels and otherworldly symbolist poetry, all I wanted was to absorb California. I frequented used bookstores and pulled tomes from the shelves of Dylan's childhood bedroom, building a library:

Mary Austin, William Vollman, Mike Davis, Carey McWilliams. The writers recounted a history that was in some ways familiar from the interior West, of frontier ambitions and pioneer failures, but also one that was unimaginably different—of the fast rise of agricultural barons and their search for an ever-cheaper workforce.

In that story of "factories in the field," the Salton Sea appeared. Like the story of the Great Salt Lake, its story was a colonial one—a story of territory-grabbing, coercive labor, and land turned toxic by chemicals. But it also diverged from the normal template of settlers finding a vast body of water and desiccating it by diverting the rivers that fed it. At the Salton Sea, it was settler greed that created the lake, nourished the lake, and then turned it toxic and left it to dry. I wanted to see it for myself.

The sea fills the base of the Salton Sink, a depression in the earth the size of Delaware. Located in the southeastern corner of California, the sink was cut initially by the movements of tectonic plates. It gradually smoothed out as the Colorado River overflowed into it once or twice every ten years for a millennium. When the river water filled the sink, it created a body of water known as Lake Cahuilla, after the Indigenous people of the region.

The Cahuilla lived in a world of sentience. "Many plants, animals, and birds were anthropomorphized, and were, therefore, part of the social universe," writes Lowell John Bean in *Mukat's People*. "They had intelligence, wishes, feelings, and . . . could act in relationship to other beings." They understood the Salton Sink to have been formed when the creators, Múkat and Témayawet, folded up the edges of the earth so that water could not spill out when it flooded. When it did, the Cahuilla moved their villages into the mountains and used harpoons and traps made of piles of stones to capture the fish that the river brought with it. When the lake shrank again, they harvested the thick layer of salt left behind on the basin floor, both for their own use and to trade.

In the mid-1800s, Anglo settlers began to arrive at the sink—
and to eye it for their colonial purposes. A geologist recognized
that it would work well as a route for the Southern Pacific Railroad; two others set up a salt mine, the New Liverpool Salt Works,
which hired Cahuilla people as its workforce, taking advantage of
their generations of experience with salt harvesting. Then, after
the 1859 California Gold Rush, Dr. Oliver Meredith Wozencraft,
one of the 300,000 white settlers lured west from the East and
Midwest, took a mule trip into the desert.

"We at last reached this—the most formidable of all deserts on
this continent," he wrote in his diary. "We found its basin filled
with turbid water." Two of his traveling companions fainted in
the extreme heat, so Wozencraft rode his mule in search of fresh
water. Eight miles away, he "reached the border of the desert and
water." It was then and there, he wrote, "that I first conceived the
idea of the reclamation of the desert."

Irrigation, he realized, would make the basin even more valuable than salt mining. Seeing the water in the sink, he had intuited
its geography—that it was lower than the bed of the Colorado
River and occasionally filled with the river's waters. The sink's
northward tilt, he recognized, made it perfect for flood irrigation,
which needed somewhere for the water to run off, while the mountain ranges surrounding it would make it a "natural hothouse," in
the words of Carey McWilliams.

Wozencraft just needed to figure out a way to get the water
there. After incorporating the Colorado River Irrigation Company and convincing the California state legislature to give the
company rights to 1,600 square miles of land, he began trying to
divert the Colorado River into the sink. But he never succeeded,
and after the bank panic of 1893 his company went bankrupt.

In response, one of the company's engineers sued for his salary.
Lacking the funds for a cash settlement, Wozencraft turned over

the company's surveying notes, maps, and engineering data. The engineer used them to form his own enterprise, the California Development Company. With the support of a wealthy financier, he negotiated a contract with the Mexican government to build an irrigation canal from the Colorado River to the Salton Sink, crossing through northern Baja California to skirt the massive Algodones sand dune fields. They started digging the Alamo Canal, a simple earthen ditch, in late 1900. Water reached the Salton Sink on May 14, 1901.

At the time, 1,500 settlers were living in the sink. To encourage more to come, the development company swapped out the basin's uninviting name for the triumphant Imperial Valley. They incorporated a second entity, the Imperial Land Company, to lure newcomers with affordable farmland that soon bloomed with Egyptian cotton and saw six cuttings of alfalfa per season. By 1904, the region had 10,000 residents.

But there was a problem. On its long path from the mountains of western Colorado to the California desert, the Colorado River filled with silt. Before long, the fine sediment clogged the Alamo canal—by then renamed the Imperial Canal—and water no longer reached the farms. The development company didn't have the machinery or the manpower to dredge it. So they rushed to dig a replacement ditch, cutting a new diversion point so hastily that they didn't bother with a headgate, the piece of metal used to regulate the flow of water.

It was a wet winter, and in the summer, snowmelt filled the Colorado and its tributaries beyond normal proportions. In November 1905, the swollen river pushed its way into the new canal with a force no one had expected, diverting not just a manageable stream but its entire flow. It coursed into the sink for a year and a half, destroying everything on the basin floor. When it finally calmed, its waters had created a 400-square-mile lake—the Salton Sea.

Ruined by the flood, the New Liverpool Salt Corporation sued the California Development Company, blaming their haphazardly built canal. The Southern Pacific Railroad also sued. The development company tried to downplay the canal's shoddy construction and nonexistent headgate, claiming that the flooding was simply a natural process and the damages inevitable. The court didn't buy it. To pay damages, the development company had to sell its landholdings to Southern Pacific. The railroad, in turn, sold the lands to the newly formed Imperial Irrigation District, or IID. Land rich, they became the new power brokers of the Salton Sink.

Over the course of the twentieth century, the influence of the Imperial Irrigation District grew steadily. During the 1920s, the irrigators pressured President Coolidge to issue an executive order designating the federal lands submerged below the Salton Sea as drainage reservoirs, officially turning the sea into a repository for the irrigation waters that overflowed off their lands—despite the fact that those federal lands formed a checkerboard, alternating every square mile, with the tribal lands of the Torres Martinez Cahuilla reservation. The federal government had promised the tribe that they would get their lands back as soon as the accidental sea evaporated. Now, capitulating to the irrigators, the same federal government ensured that the sea would stay full indefinitely.

Soon, the feds helped out the irrigation district even more. At the end of the 1920s, Congress authorized the construction of the Boulder Dam, now known as the Hoover Dam. Imperial Valley farmers argued in favor of the project, telling lawmakers that the dam was necessary both to prevent more floods and to allow them to exploit the valley to capacity. Once the dam was built, they received the largest cut of the water it stored. They had so much Colorado River water that they could waste it—except that, in the vocabulary of water law, profligate use was, in fact, beneficial. Every irrigation season, water thick with dissolved fertilizers and pesticides rolled off Imperial Valley fields and filled the Salton Sea.

Then, a second round of water flowed to the sea. After irrigating, the valley fields bloomed white with salts. Everywhere on earth besides deserts, millennia of rainfall have cleansed salt out of the soil. But in arid lands the salts remain, and they come loose when water flows through the soil. To remedy this, Imperial Valley farmers flushed their fields with more water. "What nature has taken geological eons to achieve, the leaching of salts from the root zone of plants, the irrigator undertakes to do in a matter of decades," writes Donald Worster in *Rivers of Empire*.

The Salton Sea grew and grew, and accrued more and more salts: salt from the mineral-rich Colorado River, salt from the dissolved minerals of agricultural inputs, and salts washed out of the soils of farmland. Starting in the 1930s, boosters marketed the sea as a "new Riviera." Los Angeles hoteliers built resorts with yacht harbors. The highway backed up for hours. Frank Sinatra and Rock Hudson came to party.

Meanwhile, the farmers of the Imperial Irrigation District got rich off their cheap water and the border region's plentiful labor. "When one now sees the hundreds of thousands of acres of farms in Imperial Valley producing bountiful crops of every type the year around and the homes with all the modern conveniences . . . and the hundreds of miles of paved highways and roads, it is difficult for one to visualize or appreciate what these people had to endure and overcome in conquering the desert," the irrigation district boasted in a 1960 promotional pamphlet. Carey McWilliams put it differently: writing in 1949, he called the farmers of the Imperial Irrigation District "a set of power-drunk

                                          SALT LAKES

ruthless nabobs who exploit farm labor with the same savagery that they exploit the natural resources of the valley."

I RECALLED HEARING of the sea just once before: reading the *New York Times* over breakfast at my dad and stepmother's house one Sunday, I'd seen a review of *Little Birds*, a film about two adolescent girls from the shores of the sea who get an ill-fated ride to Los Angeles. No one in my family had ever mentioned the strange lake, not even my dad, who had attended a year of college in Redlands and remained forever fascinated by the withdrawn, hunkered-down lives of California desert rats. But Dylan knew about it. That was part of why I liked him: he was such a through-and-through Californian, in its provincial, rather than glamorous, strain.

We had met three years earlier, outside the lecture hall of an art history course. I'd seen him in the class and been allured and mystified. He said he was a sophomore, but I'd never seen him before, and between the chapped skin of his face and the confident way he spoke in class, he seemed too old to be one. He was five-foot-six to my five-nine, with a square face and curly, sandy blond hair and retro country-singer glasses that were always sliding down his flat nose. His forearm was tattooed with a medieval illustration of the Dance of Death. I liked the way his clothes combined Western wear and scholarly style—he wore snap shirts and lace-up boots together with wool coats and gray scarves.

One day, I arrived early, imagining that perhaps he would, too, and that he would talk to me. It worked.

"What's your shirt?" he asked as we stood outside the empty classroom.

"It's from an organic dairy farm where I volunteered this summer," I said, pleased. I'd tried to choose something he might comment on. He was wearing a white snap shirt with square dancers embroidered on the shoulders.

"Did they milk by hand or with machines?" he asked.

"By hand," I said.

"I love milking by hand," he said. He didn't offer any information about why he had done this, but I could see from his forearm that it was true. "I like leaning into the alcove the cow makes with her hip—it's such a warm, feminine space." I blushed. He was heady and cute and knew about cows.

Three days later, I was sitting in a coffee shop reading Joan Didion's essay "Notes from a Native Daughter" when a voice asked whether it could sit down in the chair across from me. I scanned the room and realized that my table had the only free seat in the shop. I looked up only far enough to see an espresso and said yes. The espresso and disembodied voice lowered themselves to seated level. It was him. My heart pounded. I couldn't believe my luck.

We talked about Chekhov and Plato, our respective favorite authors. Dylan explained that he was a transfer student; he'd already done two years of college, followed by a year off working as a mule-packing guide. He had an accent I associated with small towns in the West, a vowel shift where e's disappear in favor of i's—"appriciate"—and the zeal for arcana and minutia that can only come from a lonely adolescence in a remote place.

"Where are you from?" I asked, desperate to know.

"Central California," he said. "Not a nice part of the state . . . the geographical center." He seemed hesitant to tell me more. "The Central Valley . . . a town called Merced . . . it's surrounded by almond orchards."

"Oh, this place." I turned the pages of my course packet and read aloud from Didion's essay: "To a stranger driving 99 in an air-conditioned car . . . the towns must seem so flat, so impoverished, as to drain the imagination . . . *I* can tell Modesto from

Merced, but I have visited there, gone to dances there; besides, there is over the main street of Modesto an arched sign which reads: 'WATER—WEALTH—CONTENTMENT—HEALTH.' There is no such sign in Merced."

We spent the afternoon together, and in the evening I brought him to a birthday party where we sat on the couch rubbing shoulders. At the end of the night, he walked me to my apartment, but I didn't invite him up. *the walk was cold*, he texted me when he got back to his dorm, chiding me. Then he sent another message recommending the Townes Van Zandt song "Rex's Blues." *y'oughta listen*, he wrote.

We began seeing each other every day. If he didn't come over to my apartment for dinner, we met for coffee in the afternoon. Sometimes we went to his dorm room and he played the guitar for me, picking out folk songs with the long nails of his right hand. He seemed immune from the fast-moving trends and disparate reference points of the 2010s hipster culture that we inhabited, so steadfast in his own interests that I imagined the world around him must feel foreign, the way it does to an elderly person who never learned to use the internet.

By November, I felt myself floating along over the grand arc that I had always imagined love entailed. I felt as though I were a foot taller and knew more about everything than anyone else. But the feeling was also more powerful than I had anticipated, so strong that it seemed it could pin me down in desperation at any moment. I knew he liked me, but I wasn't sure how much.

That December, my brother and I spent the winter break in Oakland with our aunt and uncle. Everyone but me had plans for New Year's Eve. Then I got a text message from Dylan. *hey want u here new year*, he wrote.

Days later, he picked me up at the Pleasanton BART station in his mom's RAV4. We headed southward and inland. As we turned onto I-5, the landscape suddenly became flat, and brown fields stretched as far as I could see. We turned off at the Sperry Avenue exit and drove until the shopping centers and exurban housing of Patterson gave way to a long, straight rural highway of farms, factories, and ranch houses.

"Fallow, fallow, fallow," Dylan said, as though the sight embarrassed him. He knew there was no need: I'd talked his ear off about Tolstoy and Platonov, the literature of working landscapes.

The Central Valley has an outlet to the Pacific Ocean near the city of Stockton, meaning it's not endorheic. But its southern half, below the San Joaquin River, is considered an independent watershed and a "partially endorheic" basin. Historically, the now-dry Lake Tulare collected the watershed's mountain runoff and stream water, but sometimes, in years of particularly deep snowpack, the water overflowed out of the basin, into the San Joaquin River, up to Stockton and out to sea.

The valley is the size of England and is the largest patch of soil in the world classified as Class 1: deep, level, well-drained, and suited to cultivation over a long period of time. On just 1 percent of the country's land, its massive industrial farms and the mainly immigrant workers that labor in them produce some 40 percent of the US's fruit, nuts, and other table food. Such exorbitant productivity required perhaps the most complete transformation of any ecology on earth. The valley was once an expanse of wetlands and grasslands. John Muir famously called it "the floweriest place of world I ever walked, one vast, level, even flower-bed, a sheet of flowers, a smooth sea." By now it is unrecognizable to that former self: the wetlands, including Lake Tulare, have long been drained to make fields, the Native communities murdered or displaced and erased from school curricula, the tule elk hunted nearly to extinction.

As in the Imperial Valley, this obsession with productivity has made the Central Valley toxic. Fumigation has smoked all the microbial and insect life out of the soil, so between the shocking green leaves of the grapevines and the trim almond trees, the land is a bare, empty brown. More than 200 million pounds of pesticides are applied to California crops each year, and their chemicals—arsenic, DBCP, chlorine, and other carcinogens—make their way into the groundwater and then into the valley's tap water. There are childhood cancer clusters in several valley cities.

We turned off 99 into downtown Merced. It was visibly poorer than all but a few sections of Denver. "There's Taco Bell." Dylan pointed, his tone deprecating. "There's a weird place where you can play poker." He continued the tour all the way to his parents' house, a few blocks west of downtown. "Here's this purple apartment building."

His parents lived on a street of small, stucco bungalows with a wide spectrum of upkeep. As we pulled into the driveway, I felt a strange, skipped heartbeat: it looked so much smaller than my parents' houses.

His mom, Peggy, opened the door and held out her arms to hug me. Where Dylan was small, she was tall and robust. She had long hair dyed reddish-brown and only wore black t-shirts with logos of businesses and causes she supported. She and Dylan's dad had met in San Francisco's 1980s punk scene.

"We're horrible hosts!" she said, pushing aside the door beads that led into the kitchen and opening the cabinets to reveal crackers and cookies. "You just have to help yourself to whatever you want."

In the mornings, everyone except Dylan's dad, who worked at the local high school, sat around the dining room table drinking Peet's coffee and making jokes about the morning paper. They ricocheted ditties and wordplay off of one another, jumping between different registers—affected pomp, the Idaho accent of Peggy's childhood, a voice they called "creeper"—and making every-thing they encountered into silly diminutives. Goofier and more childlike than I had ever seen him, Dylan moved loudly around the house, carrying a guitar and singing his desires. "Mumsieeee, I want to go to J. C. Penney and buy jeans, and then I want to go to J & R for tacooooos," he sang, and after the song they went and did exactly those things. Other days, we lounged in the living room in our pajamas and socks reading and watching TV. In the afternoon, the family, neighbors, and friends gathered under the pomelo tree in the overgrown backyard for "beer o'clock."

The easy conviviality of Dylan's house was something I had never experienced. I had grown up between two households that were each tense in their own way, both of my parents tempera-mental and engulfed in their own needs. I'd learned to keep quiet and to ask for very little. So when they told me to make myself at home, I did what I was used to: I slunk around finding corners to silently read in and addressing everyone with a polite formality they kindly mocked. I'd never been allowed to project myself into space the way Dylan did; I didn't know what it was like to relax at home. But, gradually, as they taught me what a warm home looked like, I began to imagine that I could have something similar myself.

Two years after that first visit to Merced, Dylan and I drove east from Los Angeles on a warm February day. It was a weekend trip that was also our first taste of the open-endedness of adult life, and we were going to explore the desert.

I kept waiting for the view to spill out into the type of open-ness I knew from the public lands in the states of the interior West. But each time I thought it would, around the bend there was a casino, or a Skechers factory, or Palm Desert. I'd thought of California as the place in Ansel Adams wall calendars—a place that was magical and stunning, protected by hardworking, erudite environmentalists—and the drive frustrated my expectations. It offered up the state as so overdeveloped and mundane that it was hard to stomach or even to pay much attention to.

We crossed the San Gabriel Valley, passing signs for Covina and Pomona and Fontana, and crossed from Los Angeles County into the Inland Empire, sprawling with distribution centers, matboard housing, and shopping plazas. Arrows pointed to exits for Ontario, Redlands, Riverside, and Jurupa Valley. I'd read about these towns, but had never imagined that they all melted together as a single glob. We pulled off at a Starbucks in Moreno Valley. It struck me as grim: tightly-packed suede houses, parking lots, and shopping plazas. But in a place that felt more like a developer's moneymaking scheme than a town, the cold tiled floors, crumb- and straw-wrapper-covered Formica tables, stale conditioned air, and mass-produced graphic design of the Starbucks felt warm and welcoming. At one table, a woman was receiving English tutoring; at another, two high-school-aged girls were playing hooky, touching each other affectionately like they were a couple.

We poured half-and-half into our burnt Americanos and continued on I-10 past Palm Springs to State Highway 62, Twenty-Nine Palms Highway. The turnoff felt like the entrance into an entirely different economy and way of life. Past the wind turbines at the exit, it was all trailers and junkyards until the town of Yucca Valley, another strip of sprawl with a Starbucks and a Home Depot—the type of place that, by virtue of being so mundane, I was learning, was where real life took place.

Joshua Tree, by contrast, seemed curated to provide visitors with the experience of rustic mysticism they wanted from the desert. We passed vintage clothing stores, hippie coffee shops, and the motel where, for an extra fee, you could stay in the room where Gram Parsons died. I thought we were going to stop and walk around. I searched on Dylan's phone for a coffee shop.

"Naw, let's just go straight to the park," Dylan objected. He told me to search instead for a trailhead. "I want to get to the hike."

"I thought we were going to see the town," I said. I felt disappointed for a moment, then swallowed it. I suspected that Joshua Tree was too cutesy for his taste, and I sympathized with his aversion to places that seemed fake.

The trail led us through molasses-colored boulders to the top of a hill. At the summit, we looked north toward a basin filled with shrubs that looked brown and crumpled after years of drought. There was another range of low mountains beyond. The landscape was different from anything I'd ever seen in the interior West. I rotated to take in the panorama. Then, through a gap in the hills, I saw the sea. It stood out in the overcast day, a flash of neon blue. I pointed it out to Dylan. He took a picture of me to send to his mom.

Back in the car, he opened the map on his phone to look at the route we'd need to take to the sea the following morning. We decided to spend the night in Indio.

As we pulled up to the city in the dark, it seemed to be little more than a built-up frontage road. We found a motel and asked at the front desk for a dinner recommendation. The smell of home-cooked Indian food wafted from the manager's apartment. The woman who came out to help pointed us back down the road. "There's a Sizzler, an El Pollo Loco, basically anything you'd want," she said.

We drove down the road to a Mexican restaurant, the one option we could find that wasn't a chain. The wait was forty minutes and the clientele all elderly white people wearing golf attire, so we left the car in the restaurant's lot and tried to walk down the street in search of something else. Dylan said that he felt like the first person to ever use the sidewalk, which lined a six-lane boulevard. We turned around.

"Fuck it, I want Taco Bell," he finally said. He loved Taco Bell; I could barely stomach it. But I didn't have a better idea. We got back in the car and headed for the point his iPhone indicated. The city's grid of boulevards was harder to navigate than it looked on the map, especially in the dark. Where were all the houses in this town, I wondered.

I took a few bites of my refried bean burrito and then set it down to busy myself reading about Indio on my phone. I knew much less about where we were than I had thought. I knew that the Coachella music festival took place nearby: the Yelp page of the motel where we were staying was full of reviews from attendees. I'd once seen a black-and-white photograph of Indio's "Date Girls," young women dressed in belly-dancing outfits as they peddled Medjools at some kind of ag expo. Dates had been imported to the region from the Arabian Peninsula starting in the 1800s, part of the process of colonizing the desert. We'd seen the plantations of palms from the highway, rising up from the desert in rectangular groups that looked like living smokestacks.

From 1940 to 2010, Indio's population grew from 2,300 to 80,000. It had expanded beyond its agricultural economy to become a bedroom community for service-sector jobs in Palm Springs and for long commutes to other jobs in the Inland Empire and beyond. Two-thirds of the city's population was Latino. In the shoreline communities closer to the sea, that figure topped 95 percent, many of them undocumented farmworkers and some members of Mexico's Indigenous Purépecha community. Much of the local housing consisted of aging manufactured homes that didn't keep out dust or heat and were prone to power outages during the extreme summer temperatures. Nearly half of the residents of the

Torres Martinez Desert Cahuilla reservation, which also bordered the sea, lived below the federal poverty line. The region had one of the highest rates of childhood asthma in the state, something many residents suspected was linked to the Salton Sea.

In the morning, we rushed out before breakfast. The road to the sea took us past more date orchards, along with grapevines springing from the gray soil, past another outpost Starbucks—this one in a gas station parking lot north of Mecca—and down the western edge of the sea. As we approached the Sonny Bono Salton Sea National Wildlife Refuge, our destination, we started to see abandoned boat rental and tackle shops. Thickets of tamarisk, also known as salt cedar, an invasive, salt-tolerant bush imported from Asia in the late 1800s for erosion control, grew up between us and the beach. We parked and walked toward the water, the fish bones crunching under our feet.

Even though they shocked me, I knew their role in the sea's story. As irrigation water flowed into the Salton Sea year after year—and especially after two back-to-back tropical storms in the 1970s—the sea became fresh enough that it could maintain tilapia and croakers, stocked by the state. Sport fishing drew locals and tourists alike. But in the summers, temperatures that languished above 100 for days on end catalyzed chemical reactions between the dissolved fertilizer nutrients and the water's hydrogen and oxygen, leaving the sea anoxic—meaning there was no oxygen for the fish to breathe through their gills. They died in mass events, turning the surface of the sea into a mess of floating carcasses. Birds died, too, sometimes faster than the wildlife refuge incinerator could dispose of them. The beach was the graveyard of years and years of this cycle. By now, tourists came simply for the feeling of eerie desolation. "The Salton Sea is a ruin porn fanatic's dream come true," the blog *Hipstercrite* had put it in 2010.

Of course, it was why I was there, too. Even as I had fallen in love with the drained, pesticide-drenched Central Valley, I hadn't expected to feel any similar fondness for the Salton Sea. I'd come to see for myself the fateful irrigation canal accident and the sacrifice zone it had birthed, ready to see a Chernobyl on the Colorado River.

Ten paces to the left of us, there was a couple taking their grandson to see birds. At the visitors' kiosk, a ranger had told us that it was a special birding week at the park. I had scoffed internally: Here? A place that kills fish? As I watched the grandfather crouch to point out an intact fish skeleton to his son, my judgment seemed confirmed.

Then, as if on cue, a snowy egret appeared, massive and majestic. It landed on a berm across the water from us and then jumped into flight again, gliding over the shallow water near the shoreline, sparkling white against the radiant blue of the water and the gray of the mountains. As we kept walking, I saw groups of shorebirds scurrying along the water's edge on their long, skinny legs. I knew the words for them—plover, sandpiper, stilt—from ornithology classes, but I couldn't distinguish them from one another. When we drove away through the Imperial Valley's enormous grid of alfalfa fields, I saw a burrowing owl, a squat, furry oval that looked like a Furby.

The birds transformed the sea from a disastrous disfigurement on the landscape into something special and redemptive. If the sea was a home for birds, I thought, perhaps it was more than a toxic scar; perhaps it was useful and necessary. I was enchanted all over again, in a way that was different from my macabre curiosity before the trip.

Later, I learned that the sea is part of the Pacific flyway, the migration route that stretches from Alaska to Argentina and historically has been used by over a billion birds each year. Over 450 migratory species stopped at the sea: geese, swans, ducks, shorebirds, warblers, orioles, buntings, and even blue-footed boobies, along with 80 percent of North America's white pelicans. It had become so critical in large part because of the draining of the Central Valley: the sea, the state's largest lake, is some of the last remaining wetland habitat left in California. The birds can spot it from miles away.

Dylan and I didn't know that the sea was about to change again. In 2003, a Supreme Court ruling had allowed a water transfer between the Imperial Irrigation District and San Diego County called the Quantification Settlement Agreement, or QSA. While the IID had Colorado River water aplenty, the city of San Diego didn't have a water source sufficient for the 2.7 million people of its metropolitan area. With the QSA, Imperial farmers' water became the supply for one-third of San Diego's water needs. It was considered a victory for water conservation: instead of going to thirsty, low-value crops like alfalfa, the water would go to people; the farmers, in turn, would be obligated to use less water-intensive practices. The problem was that far less water would make it to the Salton Sea. The sea was going to shrink.

The QSA required "mitigation water" to be sent to the sea until 2018, and directed a plan of action to be developed in the intervening time. But no single plan came together. There were too many diverging opinions about what the sea was good for, whether it was good for anything at all, and what should be done about it. When drought came and farmers had to follow new mitigation measures, the sea's prophesied decline started to become visible ahead of schedule. The water got saltier; the beach stretched further; fewer birds stopped over; dust blew off the surface.

By chance, the novel I'd brought with me on the weekend trip, Chingiz Aitmatov's *The Day Lasts More than a Hundred Years,* described a similar situation. In it, a character recalled visiting Kazakhstan's Aral Sea with an older friend. "The trip had upset [Kazangap, the old man]," the book narrated. "The sea had receded. The Aral was disappearing, drying up. They had had to walk for ten kilometers over what was once sea bed until they reached the water's edge. Here Kazangap had said, 'How much this land cost—it was at the price of the Aral Sea. Now it is drying up; one can say much the same about a man's life.'"

The Salton Sea, I gradually understood, is a paradox. Created by colonial infrastructure and reckless water use, it became something of value for birds and other creatures. But the floods of fertilizer- and manure-rich runoff that filled the warm sea made it unhealthy for the life it sustained. Maintaining the sea means letting that colonial, wasteful, poisonous status quo continue. If you conserve that water instead, you kill the sea, dry up habitat, and let toxic dust blow into the homes and lungs of local residents.

For me, coming to California was a lesson in learning to live with this kind of dissonance. I'd expected striking natural beauty and vanguard environmental consciousness; I'd found toxicity and sprawl. The world I'd come into as an adult wasn't the American West that Wallace Stegner knew, or even the one my parents knew. It was one that was damaged and poisoned, and one that I had to learn to appreciate, protect, and make a home in nonetheless.

This was the first of salt lakes' many lessons for me: places that seem ugly or desolate are vital and complex in ways that you don't notice until you give them a chance. Living in the world of my adulthood would require learning to find beauty amidst dust, bad smells, and record heat, not for their own sake but as a way of working toward something else: clean air, spectacular views, crisp mornings. Dylan's family taught me that lesson among the cloverleafs, processing plants, and contaminated water of the Central Valley, and the Salton Sea taught it to me with its smell, fish kill, and persevering bird life: it's possible—and even, in the world we live in, necessary—to make a home in an environment that is inhospitable, toxic, and mundane. Indeed, those patches of strange wildness may be the only places where it is possible.

CHAPTER 3

IMPERFECT RECOVERY

Aral Sea, Kyzylorda, Kazakhstan

A FEW WEEKS AFTER DYLAN AND I VISITED THE DESert, I opened an unexpected email. I'd received a Fulbright fellowship to conduct research in the former Soviet republic of Kyrgyzstan. Central Asia had fascinated me since childhood. I wanted to experience the region's broad steppes; I was enchanted by the idea of yurts and nomadism.

It also seemed to be the place to test my belief that studying the literature of another expansive landscape would teach me how to write about the West. Though my college education was in Russian literature, the real expansiveness was not in Russia itself but in the Central Asian countries that were its former colonies. So I'd proposed to the Fulbright Commission that I would study environmental ethics and Kyrgyz literature. I had an idea for an interdisciplinary approach, blending literary analysis and ethnography. I envisioned traveling on horseback between summer encampments and herding livestock to alpine meadows, interviewing shepherds about their pasture management, and scouring novels and epic poems for their depictions of traditional pastoralist techniques. I was particularly interested in the work of Chingiz Aitmatov, the author I'd been reading at the Salton Sea, a veterinarian turned novelist who, by writing about animals and nature, had managed to subvert the censorship of the Soviet era, hiding criticism and philosophy within what appeared to be simple, provincial socialist realism.

I also couldn't go all the way to Central Asia and not see the Aral Sea. So that spring, I arranged to visit it with Lillian, another Fulbright fellow who was spending the year in Almaty, Kazakhstan's former capital and largest city.

Spanning the border between Kazakhstan and Uzbekistan, the Aral Sea was once the world's fourth-largest lake. Its endorheic basin, more than four times the size of the Great Basin, drains the water from the mountain ranges and deserts of all the Central Asian countries: Kyrgyzstan, Afghanistan, Turkmenistan, Tajikistan, and Uzbekistan. Like other salt lakes, it is the remnant of an ancient sea, the Tethys. The word *aral* means "island" in Kazakh, and there were once over 1,100 of them in the Aral Sea, along with countless lagoons and shallow straits. The vast deltas where rivers brought water into the sea were breeding places for fish. The city of Aral, where I was headed, had a busy port that supplied one-sixth of the Soviet Union's fish, over twenty species of them. Practically everyone in town worked at the harbor. Tourists made the days-long train trip from Russia for the sea air. Then, the sea became one of the globe's most serious environmental catastrophes. Just as the decline of Great Salt Lake came about in parallel with the expansion of alfalfa cultivation in the Great Basin, the desiccation of the Aral was a result of the Soviet Union's demand for cotton.

Before it was colonized by Russia, Uzbekistan was not a nation, nor was "Uzbek" a unified ethnicity. It was a region populated by numerous small ethnic groups, some of them nomadic and some of them residing in cities that dated to the era of the Silk Road. In the 1920s, the Soviet government collectivized the people of the region, forcing nomads, landless peasants, and farmers alike to settle in gridded villages and to work side by side on the state-owned collective farms at their center. Those who did not want to be collectivized fled over the border to Afghanistan; the rest became serfs to the state.

The USSR had a centrally planned economy that was aimed at self-sufficiency. To achieve its ambitious production goals, the central government assigned each of its republics commodities that they were obligated to produce in ever-increasing quantities. Uzbekistan, Moscow decided, would grow cotton.

Producing cotton domestically had become a necessity for Russia in the 1860s, after the US Civil War disrupted global trade and the Emancipation Proclamation ended the era of cheap cotton picked by slave labor in the American South. From 1922 to 1928, cotton acreage in the Soviet Union increased from 174,000 to nearly 2.4 million acres, much of it in the Uzbek Republic. By 1960, when the Aral Sea's decline became noticeable, over 60 percent of the republic's arable land was dedicated to cotton—and production was just ramping up. In the next twenty years, the country's irrigated acreage expanded to 4 million acres and cotton production increased from 3 to 5 million metric tons—nearly all of it picked by hand, often by corvées of children, university students, and urban workers.

These ambitions required all the water the desert's rivers could give. Until 1960, Soviet water planners diverted half of the 120 billion cubic meters of water that flowed toward the Aral Sea for cotton production. By the mid-1980s, however, they had increased diversions so much that only 2 to 5 billion cubic meters reached the sea each year—even as it lost 40 billion cubic meters a year to evaporation. With its balance of inflows and evaporation so out of whack, by 1990 the sea had divided in two and its volume had shrunk by half. Its water level dropped 50 feet, its surface area decreased to less than one-tenth its original size, and its salinity rose from 10 grams per liter to over 100. No body of water in the world had ever dried up so quickly.

LILLIAN AND I had become friends after being paired as roommates during the Fulbright conference halfway through our grant period. We had a lot in common: both writers, both former Russian majors fascinated by Central Asia. She read as much as I did, and we could talk endlessly about books. Eight years older than I was, her writing career seemed established in a way I could only imagine. I hung on to her offhand comments as wisdom. We gossiped about people we knew in common and people we didn't, describing our friends to each other and analyzing their personalities over early beers. "When I find something funny, I can count on you to find it funny too," she told me soon after we met. "You always need someone you can look across the room and make eye contact with. But there aren't that many of those people in the world."

At the same time, accompanying Lillian in the world reminded me of being with my parents during my childhood: walking two steps ahead to clear anything that would set her off, a constant, painstaking routine necessary to avoid being collateral to her fury. After years of living away from my parents and being partnered with calm, trusting Dylan, I felt the tricks I'd allowed to go dormant coming back. I'd considered myself to have recovered from that kind of vigilance. But just as it had in childhood, the discomfort had an ambivalent pull. On one hand, I sensed that Lillian was asking me to care for her in ways that I had long ago stopped doing for anyone. On the other, I was proud that I had a talent for keeping her calm and buoyed.

LILLIAN AND I met at the Kyrgyz–Kazakh border to catch a shared taxi to the train station in Shu, an hour away. One of the other seats belonged to a man in a tracksuit; the driver kept us waiting for the other passenger, who had left her bags and gone off somewhere else.

The driver assured us we'd make our train. But Lillian was already in a bad mood. During the previous weeks, we'd been communicating to plan the trip. Not knowing how to get to the sea from the train stop at the town of Aral—it had receded so much that it no longer reached the shores of town—I'd booked lodging and a driver with the only tourism agency I could find that operated in the area. It was more expensive than either of us had expected, and Lillian seemed determined to let me know that she was unhappy about it. I ignored her complaints. I had made peace with overpaying—it was worth it, to me, to see the sea. She didn't have to come if she didn't want to, I told myself. Besides, I'd saved both of us money by taking a minibus to the Kazakh border two weeks earlier and buying our train tickets, instead of paying the company to book them for us.

Finally, an elderly woman arrived with more shopping bags and we got on the road. I watched Kazakhstan's irrigated fields and shaggy burros roll by. Twenty minutes later, the driver stopped the car. It was the last night of Ramadan, and the sun was setting; he had pulled over to the side of the road to pray. "Five minutes," he promised us. Lillian sighed in agitation. Finally, we made it to the train station with fifteen minutes to spare.

The platform felt like a postcard from another era. Elderly women wrapped in silk scarves peddled meat dumplings out of stacked metal steamer pots, round *lepyoshka* bread baked with black sesame seeds, and fish that hung from tall metal racks, covered in flies. When the train pulled in, it was one of the old Soviet-era ones, with a gold stripe along the side.

We were traveling second class, in an air-conditioned compartment with two pairs of bunk beds. After we left the station, the country's breadth revealed itself: brown steppe occasionally populated by villages of white or brown mud-brick houses with weeds growing from the roofs, surrounded by horses, sheep, and camels grazing on the desert grasses.

I pulled out *The Day Lasts More than a Hundred Years*, the novel I had brought to the Salton Sea. Now, I wanted to read it in Russian, from the set of complete works of Aitmatov I had bought in Bishkek. But even after years of studying Russian, the reading was slow going.

The first scene described a fox running along a lonely steppe train track—perhaps, I thought, the same track we were riding. "The dry steppe was a difficult place for plants," Aitmatov wrote. Before I got any further, I caught sight of Lillian watching me. I could tell she wanted to chat. Finally comfortable on the train, she'd become chipper.

"Have you ever been religious?" she asked.

I told her that my dad and stepmother went to church, but held back from saying anything about my own beliefs or experiences. It wasn't the first time since meeting her that I had felt like she was trying to prod me into a degree of intimacy that made me uncomfortable. Her emails referred to "financial free-fall," "emitting fallen-woman vibes," "having resolved to fuck my way through Almaty," being "on the verge of a nervous breakdown." My replies had stuck to funny anecdotes about my research and book recommendations.

One of the people sharing our compartment heard our conversation and leaned over the ledge of her top bunk. When we first entered, I'd caught sight of an Apple laptop and realized that our traveling companion must be another foreigner. She introduced herself as a veterinarian and epidemiology PhD student from England. She was traveling with a Kazakh veterinarian—the man who occupied the bunk above mine—to research the cause of the mysterious die-off of the saiga, an endangered steppe antelope with a long, funny nose. Over half of the species' population had recently died in a single weekend. She pointed to a leather jacket draped over something large between the bunks.

"That's the petrol for our expedition," she whispered. "They wouldn't let us bring it on, but he took care of it." She gestured to her traveling companion.

Five hundred miles later we pulled into Kyzylorda, a rice- and oil-producing city of 150,000 on the banks of the Syr Darya, one of the two rivers that historically provided 90 percent of the Aral Sea's water. (*Darya* means "river" in the Turkic languages.) At this last urban outpost before the Caspian Sea city of Aktau, a day and a half later, the whole train seemed to spill out, including our veterinarian bunkmates, who had arranged to rent a Jeep and drive all the way to the country's capital, Astana, over 1,200 miles. The Kazakh veterinarian donned the leather jacket that had been hiding the gasoline and lifted the huge white metal can out of the train. Fishmongers boarded and walked up and down the aisles with their metal racks, along with people selling pens and children's clothes.

Lillian pulled a pamphlet out of her backpack and slid it toward me, her eyes wide. After overhearing her curiosity about religion, the veterinarian had slipped her an evangelical tract. I was glad to have kept my mouth shut.

We still had many of our twenty-seven hours left to go. After a few of them, a huge pink sunset spread across the desert. We looked at our phones and realized that we'd crossed into another time zone. Out the window, the brown steppe had been replaced by wetlands. The train was running parallel to the Syr Darya and had followed it to its long, braided delta.

The Syr Darya flows 1,400 miles from the mountains of Kyrgyzstan to the Aral Sea. First, it descends into the Ferghana Valley, a contested border zone where Kyrgyzstan, Tajikistan, and Uzbekistan meet. There, it is diverted into the Great Ferghana Canal, which was built in 1939 by 160,000 newly collectivized Uzbeks and Tajiks who dug 172 miles by hand in 45 days. It then crosses into Uzbekistan, where it's diverted into canals that provide water to the city of Tashkent. It loops back into Kazakhstan's Kyzyl Kum—"red sand"—desert, and finally empties into the Aral Sea in the broad green marshes through which I was now passing.

The delta of the Amu Darya, meanwhile, is on the southern, Uzbek, side of the sea. The river is just 100 miles longer than the Syr, but because it is fed by glacier melt and snowfall from the Pamir and Tien Shan mountains, it historically brought twice as much water into the Aral. After descending the Tajik foothills, the river forms the border between Uzbekistan and Afghanistan. There, a quarter of its flow is diverted into Turkmenistan through the 850-mile Kara Kum—"black sand"—Canal. The canal's 1958 construction opened up huge areas to cotton monoculture, and provided a water supply to Turkmenistan's capital city of Ashgabat. After that, the Amu–Bukhara Canal, built in 1963, diverts water to the city of Bukhara, Uzbekistan, and its surrounding farmland. Finally, the river comes to its most ancient diversion: a two-thousand-year-old complex of canals leading toward the city of Khiva. Then, somewhere before it reaches its delta, it dries up.

When, exactly, the river stopped reaching the sea is unknown: the Soviet government didn't publicize it. Perhaps they didn't even take note. We do know that the industrial fisheries collapsed in

1980, portending the region's economic decline. The sea split into north and south lobes in 1986. Things were getting dire just as the Soviet Union was breaking up; its crumbling government was not going to deal with the crisis, and the young governments of newly independent Kazakhstan and Uzbekistan were stretched thin running their countries. So, things kept getting worse. The sea's south lobe split again, into east and west lobes, in 2003. The very week I arrived in September 2014, satellite imagery from NASA showed that the eastern lobe had disappeared completely.

After religion, Lillian tried a new topic. She asked whether I'd ever slept with a woman. No, I told her.

"Didn't you ever want to?" she asked.

I did, but I didn't know why. I knew that I'd always felt envious of my queer friends. I wanted to be part of their club of being at odds with the world, and being affectionate while doing it. At lunch in high school, I'd often sat with the gay "scene kids," whose early-2000s style of dyed black hair with asymmetrical bangs, square glasses, and checkered Vans seemed so impossibly cool that I felt grateful each time they let me join them. But I was too shy to take them up on the invitation to go to their gay-straight alliance meetings, even when the club's presidents became my close friends.

During college, I imagined that maybe one of my female friends and I would sleep together, simply to satisfy the curiosity I believed everyone had. But I slowly realized that most of them didn't share the desire to find out what it would be like. I wasn't willing to take the risk of wondering aloud about my sexuality, and I never understood how to intuit that someone was interested in me, or attracted to me; my assumption was always the opposite.

So, I couldn't tell if Lillian was trying to provoke me into confession or propositioning me. Once again, I avoided answering the question with my own thoughts. I told her about how most of my male partners had been bisexual. It hadn't been a conscious decision, but a repeated happenstance that at a certain point had stopped surprising me. I liked how they could shift between gender presentations, their virile exterior occasionally baring a more feminine sensitivity beneath, and how they treated masculinity as an aesthetic endeavor, achieved through attention to tartans and thread counts. They made sex communicative and playful in a way that put me, who didn't always like sex that much, at ease.

Finally, the train pulled into Aral. A driver from the tour company took us to our guesthouse. There were camels in the streets. Even after midnight, it was sweltering. At the house, the hostess showed us to a simple room with two twin beds and a fan. I watched Lillian's mood change. I turned on the fan, asked her opinion about opening the window, offered her my pillow, and finally went to bed, worried that if I said too much or made a wrong move, she would yell at me.

When I woke up in the morning, Lillian had gone for a run. I chatted with the hostess, a woman who appeared to be in her sixties or seventies, while we waited for her to return. When I told her that I had been living in Kyrgyzstan, her face fell in pity. It surprised me: the desolation of Aral seemed to deserve much more sympathy than lush Kyrgyzstan. But she still thought of her town as a prosperous coastal port. "Everything is good *here*," she said, insinuating the relative poverty of Kyrgyzstan in comparison to its northern neighbor. I asked her what kinds of jobs people had in Aral. She shrugged. "Doctors, teachers, fishing."

Never mind the toxic dust, I thought cynically, the fertilizer and chemical weapons particulate in the air, the unemployment rate long the highest in the country, the infant mortality rate quadrupling, the warning to women not to breastfeed their babies, the epidemic tuberculosis and throat cancer, the rates of typhoid eight times the national average, the child deaths from respiratory failure, the anemia, brucellosis, asthma, heart disease, kidney disease, and illnesses of the nervous system. Everything is good *here*.

When Lillian finally came back, we ate breakfast and set off for the sea. With daylight, I could see the town. The buildings matched the ground, with a color scheme that ranged from sand to concrete gray, plus a handful of fading pastels. Many of the buildings constructed since the end of the Soviet Union were made of shipping containers. Most houses had wide platforms in front of them, with families lined up, asleep; as an adaptation to the heat, Aral was largely nocturnal. It was also noticeably empty: between 1980 and 2000, forty-five thousand people had left the region.

"Weren't you worried about me?" Lillian asked as we drove.

"I figured you would make it back," I said calmly. "I wasn't sure how long of a run you went on."

"I was gone a long time. I got lost. *I* was worried," she said.

"What could I have done?" I asked calmly.

"You could have told the hostess you were worried," she said, slightly more aggressively. I nodded silently. I had spent the morning trying to distract the hostess from being offended by Lillian keeping breakfast waiting, and then invisibly ameliorating her brusqueness once she did return. I stayed quiet, waiting for Lillian to calm down.

The sea was an hour's drive on a two-track road through the former seabed turned desert, whose ground was the color of sand but clumpy like dried mud. But like the Salton Sea, it wasn't empty of life. There were camels grazing on saltgrass that grew from the polygon shapes of the dry ground. People had lived in the Aral Sea basin for thousands of years. They say the Karakalpak women who inhabit the Uzbek side of the lake knew two hundred ways to cook fish. I started to understand our host's confidence that things were good here.

Our first stop was at some old, abandoned ships, now many kilometers from water, just like the village adjacent to them. Locals had been taking them apart for their scrap metal, and only a few remained. Two were small fishing boats, their interiors littered with rocks and thick broken glass. Their blue paint was wearing off and being replaced by graffiti. We climbed in and out of them, imagining their past seafaring. Lillian took a photo of me standing in the windowed bow, where the thin metal of the floor was rusted through. Someone had scratched *jilu*, the Kazakh word for warmth, into the exterior. The third ship was a barge, three stories high, with huge component parts scattered around it. It was covered in wheat-paste graffiti of sailors—some reclining, some drinking, some despairing. One slept shirtless in a hammock; another wore a captain's uniform. One dressed in boots, loose pants, a coat and a knit hat bore the name Makai.

A herd of camels had stationed themselves in the shade cast by the barge. Some had curly, feminine hair on their heads. Many wore large wooden necklaces. They were a pleasant surprise of the region: though they figured prominently in *The Day Lasts More than a Hundred Years*, I had never imagined that I'd see them in such quantities—or that the desiccated seabed could provide them enough sustenance to survive.

We circled the camels, crunching through the layers of salt and dust. This desert where the Aral Sea had once been is known as Aral Kum—"Aral sand." Up to a hundred million tons of salts and dust blow off the Aral Kum each year, and the impact is felt across the world: the dust coats glaciers from the Himalayas to Greenland with a gray film, accelerating their melt, and it has been found in the blood of penguins in Antarctica.

Like those of the Great Salt Lake and the Salton Sea lakebeds, the dust particles are also toxic. They carry pesticides like the defoliant Butofos, which was used to make cotton plants drop their leaves, for easier picking of the bolls, until it sickened and killed so many workers that its use was discontinued and the remaining chemicals, left in barrels, seeped into groundwater. There's also DDT, nitrogen fertilizer, hazardous industrial wastes, and anthrax: Vozrozhdeniya Island, in the middle of the Aral Sea, was a secret facility for the USSR's biological warfare program from 1954 to 1992, and there were open-air tests of anthrax bomblets. Toward the end of the Soviet era, the regime transferred hundreds of tons of the substance from other facilities to the isolated island, where it was buried in drums and in simple sand pits, mixed with bleach.

We continued to the sea, kicking up more of our own dust, but also moving into wetter, more fertile ground. We started to see green reeds, and by the time we reached the shore, red and green plants sprouted thickly in front of us—some short grass, some rushes rising higher than my waist. Lillian and the driver lit cigarettes, finding little to look at. I walked toward the water, also trying to find something to see. How do you appreciate something as vast as an inland sea from one tiny point on its shore, I wondered.

I dipped my hand into the marshy water, then tasted a salty bead off my finger. "It just looks like a marsh, nothing strange or apocalyptic," I wrote in an email to a friend later. We headed back into town, passing a community of shipping-container houses whose tin roofs were all painted a sturdy and inviting deep blue. "You can build one of those for four hundred dollars," the otherwise taciturn driver said.

THAT AFTERNOON, we went to the local fruit market to meet up with an anthropology PhD student from the US who was conducting research in Aral. She was elegant, with long black hair in a braid down her back. She spoke Kazakh and knew much more about the region than we did. She wanted to visit a salt factory twenty kilometers outside of town. We found a taxi willing to take us there. We drove on a dirt road, passing desert, sheep, and cemeteries before finally approaching a sign that marked the town limits of Aral Tuz— "Aral Salt." The anthropologist asked the driver in Kazakh whether everyone in the town worked at the plant. "Not everyone," he replied. "Some just have livestock."

We pulled up alongside the factory and the huge mountains of salt that sat just beyond it. The front of the taxi was splattered highlighter yellow from hitting dragonflies. We noticed an open door and walked over to look in. Inside was a room of parallel conveyor belts, each serviced by women in white dresses and white, nurse-like caps. Some were putting the salt into plastic bags; others were picking up the filled bags and tying them closed. The four of us, the taxi driver included, stood in the doorway for several minutes, mesmerized. Finally, we watched as one of the women removed herself from her conveyor post and walked to the security guard in the corner. Soon, a casually dressed man, perhaps the floor manager, came to the doorway and told us we had to leave. We obeyed languidly, trying to take in the strange scene—a tiny corner of productivity in the middle of the desolate, quiet landscape—for as long as possible.

The following day, we met up with the anthropologist again. There was something else she wanted to see: the Kok-Aral dam, the Kazakh government's effort to revive the northern Aral. In 2005, the Kazakh government had built the dam from Kok-Aral—a former island that had become an isthmus as the sea's water levels had dropped—to a point just east of where the Syr Darya emptied into the lakebed. The dam barricaded water from the Syr Darya, preventing it from flowing southward to the Uzbek side of the sea; by doing so, it had allowed a relatively small northern lobe, some 1,800 square miles entirely on the Kazakh side, to regain a degree of health. In addition, the government began helping farmers improve their irrigation systems so that they used less water. The amount of water reaching the lobe doubled and the water level shot up, allowing the little Aral to reach its target depth of 42 meters—about 138 feet—just a year after the dam's completion. Anemia rates dropped. More than five thousand people moved back to the region. The fish catch increased tenfold, and native species started to return, along with migratory birds like flamingos and pelicans. The northern lobe was in recovery.

As we turned toward the dam, passing through the delta villages, more and more green appeared in the landscape, which meant more and more cows and camels grazing. On a bridge over a tributary whose banks were thick with green plants, the anthropologist struck up a conversation with the fishermen leaning over the bridge's railing. "The fishing is good!" they said.

Before the dam was built, the sea was one hundred kilometers from the city; by the time we visited, it was only twelve. There were talks of raising the dam or building a second, which would bring up the water level another six meters, allowing the sea to reach the town of Aral in seventeen years. Residents imagined the days when their city would again be full of people selling fish that were abundant to catch and cheap to buy. The project's chief engineer told the *New Internationalist*, "In the eyes of the locals, [the dam] will only be a total success when the water is running right up to Aralsk's port again."

In human health, the word *recovery* implies a return to the state that preceded illness or injury. We use it in serious and prolonged situations, like cancer or a life-threatening car accident. A full recovery is one in which the traces of ailment are erased. The completeness implied is perhaps clearest in its metaphorical use

in sports: in tennis, for instance, the "recovery step" is the movement that brings a player back to the starting position, prepared
for the opponent's next shot. Athletes run drills to make their
recovery steps as imperceptible as possible.

In reality, recovery is more complicated. "The probably forever
of never-all-better," poet Anne Boyer calls it in *The Undying*, her
book about recovery from breast cancer. Ailment impacts us. So
does the process of recovery. "That 2-million-volt monster is my
only ally in the major battle—but an awesome and terrible ally,
for even while it is killing the cancer I know what it is doing to
me," wrote Rachel Carson of the treatment she received for her
own breast cancer, from which she eventually died. Coming back
to health is arduous and imperfect, much more than a smooth
shift of body weight.

"Recovery" is also a key word in the lexicon of the century-old
addiction support organization Alcoholics Anonymous. Inspired
by the Christian spirituality of psychoanalyst Carl Jung, the texts
of AA teach recovery as a spiritual process more than a medical
one, focusing not only on abstaining from alcohol but also on
mending one's relationships and sense of self.

The twelve steps that comprise AA's understanding of recovery
begin with the idea of a higher power—however the alcoholic con-
ceives of such a being, but called "God" as a shared shorthand.
"We admitted we were powerless over alcohol—that our lives had
become unmanageable," reads step one. "Came to believe that
a Power greater than ourselves could restore us to sanity," step
three; step four: "Made a decision to turn our will and our lives
over to the care of God as we understood him."

In the cosmology of the twelve steps, recovery is a process
of submission to the will of God, of prayer and meditation,
and of humility. Step ten is to continually "take personal inven-

                                                          SALT LAKES

tory" and to promptly admit being wrong. So, recovery never ends. Alcoholics who stop drinking are forever "in recovery"—never "recovered." They can't go back to drinking, stop believing in their higher power, or stop articulating their flaws. "The AA member has to conform to the principles of recovery. His life actually depends on obedience to spiritual principles," reads the "Big Book" of Alcoholics Anonymous. The directives can seem stark and dogmatic, but they are also salves. They offer a framework for enduring the difficult longness of life—and the feeling, to those abstaining from an addiction, that that life has been tremendously reduced.

At the sea, too, recovery comes with reduction. The Aral's additional dam was never built. Construction in Kazakhstan never received Uzbekistan's blessing. Instead, many residents are now raising concerns about the soundness of the original dam. Over twenty years old, they worry it will fail—and never be replaced. Even with an additional dam, it would never match the sea of residents' memories and inherited stories. The basin's waterways are too fundamentally changed to allow a simple reversal of fortunes.

Utah scientist Wayne Wurtsbaugh refers to using dams to reduce the size of saline lakes as "the Aral Sea solution." That kind of intervention is now on the table for Great Salt Lake. But while, at Aral, it made fishermen and other locals happy to see some piece of the sea return, the solution is imperfect. Perhaps most importantly, it doesn't solve the problem of dust storms. The rest of the Aral Sea—the 95 percent of the bed not covered by water—remains one of the world's most severe ecological disasters. The Aral will never be the same again. As in twelve-step cosmology, the Aral Sea's recovery was powerful, but also elusive and partial.

W E   D R O V E   B A C K to Aral and headed for the train station. I admired the grazing camels along the road for a last time. On our return train trip, Lillian and I traveled in a third-class wagon full of bunk beds. It was sweltering. I woke up in the middle of the night to flip over my pillow, which was soaked with sweat. At some point in the night, Lillian had stripped down to her underwear and I had thrown off my shirt; the conductor had covered each of us with Kazakh Railway-issue sheets for modesty.

By morning, the chocolate bar Lillian and I brought with us had melted on the table between our bunks. I was stickier and more uncomfortable than I had ever been, but mostly I was worried about her mood. I made jokes, trying to keep anger from finding its way in until I got off the train at Shu and left Lillian to continue to Almaty on her own.

My fellowship was ending; Dylan came to visit the following month, and then I returned to the US. Lillian wrote me shortly after I touched down in Denver. "I've progressed to burning myself with cigarettes," the email said.

My skin prickled. I sensed that I was being baited, again, into an intimacy I hadn't agreed to. I replied distantly, certain that I wasn't offering her what she wanted of me, but also cognizant that I had never offered more than I could give. A day later, she replied that she was finished with our friendship. "I've had enough of your stiff upper lip," she wrote.

I considered replying with an apologetic appeal to save the friendship. But something bristled deep inside of me. I was tired of the vigilance it required. I archived the message and turned my phone face down on the table in front of me. Partial though it may have been, I wanted my recovery back.

II.
RECALIBRATIONS

CHAPTER 4

THE COMMON GOOD

Mono Lake, California, USA

Fʀᴏᴍ Dᴇɴᴠᴇʀ, Dʏʟᴀɴ ᴀɴᴅ I ʜᴀᴅ ᴘʟᴀɴɴᴇᴅ ᴛᴏ ꜱᴘᴇɴᴅ the rest of the summer at his parents' house in Merced. I sent a friend an email about the road trip we planned to take to California. "We bought a tent and we're going to camp in the Great Basin National Park in NV and then at Mono Lake in CA!" I wrote, my excitement unbridled.

Not being a Californian, I'd only heard of the lake from the ʟᴏɴɢ ʟɪᴠᴇ ᴍᴏɴᴏ ʟᴀᴋᴇ bumper sticker on the minivan that belonged to Dylan's childhood best friend's family. I didn't know what it was like to wake up to its sunrises on camping trips, explore its crenellated tufa towers, or eat fish tacos at the Lee Vining Mobil station.

The first night, we camped in the Great Basin as planned. It was remote and alive: on the park road, we had to drive slowly so that we wouldn't kill any of the dozens of jackrabbits crossing in front of us. Our windshield was splattered with dead insects. We pitched our tent in the dark and listened to kangaroo rats scurry around the sandy soil.

In the morning, we crossed the state on Highway 50 and turned left at the military-base town of Fallon. As we drove south, a mountain range rose up to our right, and suddenly the valley to our left filled with water, the guardrail protecting us from a sheer drop. A reservoir, I figured, until we saw signs for Walker Lake. A salt lake. We kept going. Eventually, the road formed a T and we headed to the right. Just before the turn, there was a massive, glistening saltpan on the left-hand side, along with rotting wooden installations from its past as a mine. Another salt lake, or at least the remnant of one.

The road went up and down, crossing the basin and range at a diagonal, and flattened out as we crossed the cattleguard marked "Welcome to California." The landscape's color scheme was changing, too, from dark purple mountains and hunter-green shrubs to bright greens and yellows accenting the pines. Even the sky seemed to brighten, playing along with the palette.

Then, I saw Mono out the car window. It looked otherworldly: a circle of blue shining bright and reflective amid the folds of the landscape. I'd never seen anything like it, even other salt lakes.

In 1872, Mark Twain called Mono "the loneliest tenant of the loneliest spot on earth." The lake occupies a crater-like bowl that is twelve miles in diameter and 150 feet at its deepest. Its surface is pierced by tufa towers—cream-colored pinnacles of calcium carbonate that are formed by both geologic processes and brine fly secretions.

The Mono Basin sits just to the north of the Owens Valley, the largest basin of the Great Basin region, and the wettest—since it's located immediately to the east of the Sierra Nevada, all the snow-melt from the range's eastern face flows to its floor, a catchment area of three thousand square miles. Local Paiute tribes know the Owens Valley as Payahuunadü, or "land of flowing waters."

Like other salt lakes, Mono was once much larger. One hundred twenty thousand years ago, it was so big that it spilled south, over a mountain range, into the Owens Valley, whose waters in turn spilled into the neighboring Searles Valley. It shrank to its

   SALT LAKES

current size around nine thousand years ago. It's a triple-water salt lake: saline, alkaline, and sulfurous. It also contains fluoride, boron, arsenic, uranium, and plutonium. Some scientists believe that because its waters are so old, they may resemble the composition of pre-Cambrian oceans more than anything else on earth.

Traditionally, the Paiute people who lived in the Mono, Owens, and other nearby basins used willows from creekbeds for spears, basket-weaving, building materials, and latticed fish traps. They hit the branches of piñon trees with sticks to loosen the nuts, which they would eat raw, roast in their shells, dehydrate and grind into flour, or boil into a rich, gravy-like soup. They used juniper wood for bows and alder bark for dyes, but avoided felling the trees so that they could harvest again and again for generations. They collected edible seedpods from water lilies and shook the slender shoots of lovegrass and wild rye to collect their tiny seeds. They burned patches of brush in the eastern Sierra foothills and broadcast the seeds of *Chenopodium*—the plant Mexicans know as *huauzontle*—and *Mentzelia*, whose seeds they ground into a substance similar to peanut butter. They set brush fires in early fall to increase the number of wild onions, elderberries, and caterpillars, and to drive rabbits into nets to shoot them with bows and arrows. They diverted the rivers to water stands of *Dichelostemma*, or blue dick, to harvest its bulb-like stems. For dessert, they collected the honeydew deposited by aphids on cattails and tule stalks and made them into sweet balls.

The tribe that lived along Mono Lake's shoreline was the Kootzaduka'a Paiute, whose name means "fly larvae eaters." (*Mono* means "fly" in the language of the neighboring Yokuts tribe.) Like other Paiute tribes, the Kootzaduka'a harvested piñon nuts, hunted rabbits and antelope, and ate seeds and berries. But the lake and its creatures played a special role in their gastronomy. They caught gulls and grebes, collected gulls' eggs, and feasted on alkali flies. They sieved the fly pupae—which grow inside hard casings like butterfly cocoons—from the water with woven baskets and laid them out to dry before taking them out of their sheaths. "A yellowish kernel remains, like a small yellow grain of rice. This is oily, very nutritious, and not unpleasant to the taste," wrote botanist William Brewer in the 1860s.

Like most other California tribes, the Kootzaduka'a Paiute were never federally recognized or allotted a reservation. (The tribe is now in the process of seeking federal recognition.) They also never ceded their lands through a treaty. Though the Native residents had explained to white settlers that they could pass through the valley as long as they did not "sit down on the grass patches," the area's potential for mining was so great that the settlers and their army were not going to comply. Across the Great Basin, Paiute tribes were dispossessed of their land first by violence, and second by bureaucracy.

After the United States claimed California from Mexico with the 1848 Treaty of Guadalupe Hidalgo, Congress passed an act stating that all existing ownership claims to the country's newly acquired lands, including Native claims, had to be presented to commissioners within two years for confirmation. Otherwise, the lands would become part of the public domain. Though they didn't publicize this rule to the Native people that had called the regions home since time immemorial, two years later, like clockwork, Congress passed another act placing all unclaimed California lands into the public domain. Federal surveyors showed up five years later and divided up the tribes' homelands into 160-acre rectangles whose coordinates were put on file to be obtained by settlers, either purchased for $1.25 per acre or applied for under the Homestead Act and Desert Lands Act.

In the Mono Basin, the most popular tracts were those along

Rush Creek—Mono Lake's largest source of water—and its tributaries, Parker and Walker creeks. According to the terms of the acts, homesteaders had to "improve" the lands they were granted by building structures, farming or ranching, and setting up irrigation-works. All of these improvements threatened Paiute livelihoods. (Just ten Paiute families were able to obtain homesteads.) The settlers diverted the creeks' water into irrigation ditches and dug wells to tap into the groundwater table. For lumber, they cut down large swaths of the piñon trees that Native residents had tended for generations. Entrepreneurs went door to door at mining camps selling the gulls' eggs and "Mono Lake ducks" that provided food for the Kootzaduka'a. The Paiute responded with death threats, and, as in Utah, the settlers couldn't understand why. "What has given rise to this sentiment, people are at a loss to conceive, unless it be the chronic jealousy of the redskins . . . contrasting the difference between their own debased condition with the continually increasing comforts and luxuries of the white settlers," wrote one.

In response to the threat of violence from the tribes, the federal government set up a military outpost in the valley called Camp Independence. It was the garrison from which they would fight the Owens Valley Indian War. Concurrent with the Civil War, the US Army carried out a war against the Paiutes and Shoshone in order to take the region for settlers. The invasion was no easy task, since the Native residents knew the landscape well and used its openness to see the troops coming from far away and hide. So the army provoked them into battle by cutting off their supply of food. Colonel George S. Evans wrote in a July 1861 letter:

These Indians subsist at this season of the year entirely upon the grass seeds and nuts gathered in the valley from the lake up, and the

The troops ambushed the Paiutes while they were gathering fly larvae, killing civilians and taking hostages, and destroyed other food sources like nuts, seeds, and grasses. By 1863, dozens of Native people had been killed, and nearly a thousand displaced to a newly created reservation at Fort Tejon.

It's so unpleasant to imagine that contemporary activities like hiking, climbing, and just driving through beautiful scenery were made possible by violence that few think about it. But it's important to know these histories, because they are the foundations of the lives we now live in the region. Understanding what happened is the first step toward recompense.

After the Indian Wars, settlers repeated the violence they had committed against Indigenous people against the waters. The water wars began when some settlers in the Mono Basin wanted to amass more land and wealth than their homestead allotments would allow. Taking advantage of the fact that California's laws were still in flux, they angled to monopolize the basin's water rights. In 1913, one landowner sued to gain complete control of Rush Creek and Parker Creek. Two years later, he sold those lands and their water rights to another settler, J. S. Cain. Cain had made a fortune in the local gold and silver mines, partially by building a hydroelectric plant to serve his operations. Now, he planned to build plants on all three of the creeks that led to Mono, and eventually to build a tunnel to send water from the Mono Basin into the Owens Valley, where he had purchased a hydroelectric plant in the Owens River Gorge.

But he wasn't the only one with that plan. At the same time as Cain was machinating, the growing city of Los Angeles needed water. Fred Eaton, a forty-niner's son who was both the former head of the Los Angeles Water Company and a former mayor of the city, knew where to get it. He approached William Mulholland, the new head of the water company, which was now a municipal utility. Years earlier, Eaton had given Muholland his first job there, as a deputy zanjero, or ditch overseer. Mulholland had no formal training as an engineer, but he had risen through the ranks to become Eaton's successor. He hadn't forgotten his mentor.

Eaton wanted to tour the Owens Valley with Mulholland. In 1904, they traveled there by horse-drawn wagon, observing the region's apple orchards and wheat fields along with its unplowed expanses of sage and saltbush. Eaton had purchased a ranch with forty miles of Owens River frontage and wanted to sell the water rights to Mulholland's utility. Mulholland agreed; LA would buy the rights from Eaton and build a reservoir on his ranch. Stopping the water there and diverting it into an aqueduct headed for the city would put downstream ranchers out of business, but this didn't bother Eaton: they would just be forced to sell him their cattle.

Eaton and Mulholland kept the plan secret until July 1905, when the *Los Angeles Daily Times* printed an article headlined "Titanic Project to Give City a River." "Three months ago, Eaton bought the holdings of the Rickey Cattle Company, comprising about 50,000 acres of water-bearing land," the story read. "It was then thought that Eaton was going into the stock-raising business here, but it has since been learned that he was securing options for Los Angeles city." Nevertheless, the paper assured, "the price paid for many of the ranches is three or four times what the owners ever expected to sell them for. Everybody in the valley has money, and everyone is happy."

Of course, that wasn't true. The local congressman called it a "government by strangulation." But Mulholland was emboldened. He moved to acquire more and more of the Owens Valley's water rights. Los Angeles voters passed a bond to finance his water project in 1907, while the Department of the Interior did him the favor of removing half a million acres of the valley from settlement, meaning that would-be farmers and ranchers could no longer apply for tracts there under the Homestead Act.

Mulholland purchased five quarries and a cement plant and built the 233-mile Los Angeles Aqueduct, diverting the water that flowed into the Owens River—enough to flood 300,000 football fields a foot deep—into a long, straight, concrete canal that would ferry it southward. Local ranchers—white settlers who had arrived to homestead before the big-city executives showed up—retaliated by dynamiting the aqueduct and holding its gates hostage for days to divert the water back into the valley. But they were no match for the powerful agency. The aqueduct began operations in 1913, with an inauguration ceremony at which Mulholland gestured to the crowds and proclaimed, "There it is. Take it." By 1933, the city owned 95 percent of the Owens Valley's farmland.

While the leaders of the Los Angeles Department of Water and Power, as Mulholland's company-turned-utility was now known, drained the Owens, they also began to eye the Mono Basin. They started buying land and water rights there in 1912. In 1919, they struck a secret agreement with the director of the Bureau of Reclamation—the federal agency responsible for building dams and power plants to encourage settlement and economic development in the West—in which the Department of Water and Power paid the Bureau to secretly prepare surveys, plans, and a draft budget for an extension of the Los Angeles Aqueduct that would drain the Mono Basin. The director of the Bureau of Reclamation lost his job over the deal, but not before he had withdrawn all public lands in the basin from any new settlement or mining claims. That made Cain the only obstacle left in the city's way. The

Department of Water and Power tried to use eminent domain to wrest Cain's Owens Gorge power plant from him by force, but lost the fight in a court of appeals.

Finally, in 1933, the agency bought him out for $7 million. It got to work building the eleven-mile tunnel that would transfer the water from the Mono Basin into the Owens Valley and eventually into the Los Angeles Aqueduct. The project was slowed down by "steam, hot water, volcanic gases, and ground caving," as scientist David Winkler later wrote, but the city persisted. In April 1941, the headgates opened. The rivers and creeks that once flowed into Mono were now channeled into the concrete aqueduct, destined for the Southland.

During the 1960s, the city added a second barrel to the system, which nearly doubled the amount of water they could divert from the Mono Basin. At the lake, the effects soon became visible. Deep-rooted plants began dying, the water level fell, springs dried up, and white dust storms exploded on land that had once been covered by water. The alkali flies disappeared, and the Kootzaduka'a people stopped visiting the lake to harvest them. By the 1970s, the relatively small group of people who knew and cared about the lake were worried.

ONE OF THE WORRIED FEW was Tim Such. As a junior at the University of California, Berkeley, Such had taken an environmental studies course for which he was assigned to write a final paper about an environmental problem that wasn't addressed by current law. He picked Mono Lake. It was an easy decision: on family visits to Yosemite, he'd seen how its water level was declining because of the Los Angeles Aqueduct, and it didn't seem like anyone was doing anything about it.

Such thought there would be a simple fix—a water law to be

applied, or an agency that could step in. But he soon learned that California law gave full ownership, not just usage rights, to its water rights holders. That meant that the City of Los Angeles could do whatever it wanted with the streams it diverted away from Mono: the water was its property.

Such brought his term paper to the Sierra Club's Mono Lake Task Force. They told him it was too goliath an undertaking. "They thought you couldn't fight Los Angeles," Such recalled in the book *Storm over Mono.* "They said, 'If you find a good legal theory, come back.'"

So he did. After graduating, he took a job as a private investigator as a way to support his real work: saving Mono Lake. He devoted his free time to mining Berkeley and Stanford's law libraries for feasible avenues. He read the California constitution, international migratory bird treaties, and the history of LADWP's Mono Extension. Then, one day, he picked up a back issue of the *Michigan Law Review* and began reading an article from 1970 by a law professor named Joseph Sax.

The article focused on the public trust doctrine, a concept that obligates states to manage bodies of water within their boundaries for the common good. Its origins date to antiquity. In the year 535, Roman emperor Justinian wrote into law that access to the shores of the Mediterranean was public, decreeing: "By the law of nature these things are common to mankind—the air, running water, the sea and consequently the shores of the sea." Later, the English adapted it into their common law, stating that the sovereign had ownership of all navigable waterways and the lands beneath them as the "trustee of a public trust for the benefit of the people."

After multiple disputes over the ownership of streambeds in the nineteenth century, the US Supreme Court brought the public trust doctrine into US law, solidifying around 1845 the idea that streambeds belonged to states. In 1892, they applied it to a case regarding the Chicago lakefront. It was a historic decision because it not only marked the application of the idea of public trust to an inland body of water, but also stated that the lakebed could not be bought and sold. That meant that as the twentieth century began, submerged lands were firmly public.

Since the Roman era, Sax wrote, the public trust doctrine had been thought of as useful for protecting waterways' utility for commerce. But, he argued, the doctrine could be applied to a much wider range of public needs. He surveyed its past applications across the country, and showed that they included uses for recreation and ecological values.

"Of all the concepts known to American law, only the Public Trust Doctrine seems to have the breadth and substantive content which might make it useful as a tool of general application for citizens seeking to develop a comprehensive legal approach to resource management problems," he wrote. The state of California had codified his argument into law the very next year, expanding its definition of public trust from navigation and commerce to include habitat and aesthetic values. Could it apply to Mono?, Such wondered. There was a precedent in state case law—and it was just a few years old.

He took the idea to the Sierra Club and other organizations. They were still hesitant to take on the behemoth Department of Water and Power. But one person listened: Tony Rossman, special counsel to Inyo County, which covers the northern half of the Owens Valley. Long dispossessed of its land and water, the county was looking for a way to take on the big-city water agency. But they would need ecological data to back up their argument and build a case.

As luck would have it, right then, that data was being collected. In the fall of 1975, a Stanford student named Jeff Burch got a flyer in the mail from his former high school biology teacher. It advertised a National Science Foundation program called Student-Originated Studies, which offered grants of $20,000 for groups of undergraduates to do original research projects. Burch was intrigued. Like Such, he had visited Mono Lake and was worried about it. He rounded up some friends—students from Stanford, various University of California campuses, and Earlham College—and reached out to David Gaines, a UC–Davis master's student who had written a report about Mono for the Nature Conservancy a year before, for advice. They proposed to spend the summer conducting a thorough study of the lake's ecosystem. Nearly all of the research about the lake up to that point had focused on its geology, meaning that relatively little was known about the life forms that inhabited it and how the decline in the water level was affecting them.

They got the grant. They spent the summer camping on a ranch near the lake's northwest corner, where they "sang, recited verse, lived largely on granola, beans, rice, and were known to take in other nonstandard substances," as John Hart writes in *Storm over Mono*. But they spent the majority of their time conducting the research that would provide a baseline understanding about the ecosystems of saline lakes, and the effects of water diversion on them, for decades to come.

Each of the group's members took responsibility for a different aspect of the research: plants, rocks, bugs, birds. They did their research on foot, using canoes and a "beat-up old boat that we got from Fish and Game," as David Winkler, one of the group's members, told me. As they traversed the lake and got to know its nooks and crannies, they gave names to the features: Pancake,

Twain, Muir, Java, Little Norway. Then, they pieced together the lake's food chain—and how the increase in salinity was affecting it.

People often refer to salt lakes as "dead," implying that they are absent of life. But the lakes overflow with life—just in specific, small forms. In the spring, Mono turns green with millions of tons of algae. Then, thousands of brine shrimp hatch from cysts into nauplii, similar to tadpoles. By mid-May, the year's generation of adult shrimp is grazing on the algae. The shrimp are extremely abundant: up to three thousand of them live in each cubic meter of water. By the end of June, their foraging turns the lake a clear blue. Meanwhile, up to 100 million fly larvae hatch from eggs laid on the surface of the lake, so many that visitors mistake them for a ring of tar. That abundance of shrimp and flies draws birds to salt lakes.

Dave Herbst and Gayle Dana were in charge of studying the brine shrimp. As fully-grown adults, brine shrimp are just one-half to three-quarters of an inch long. They're among the most primitive creatures on earth; they're thought to have arrived at Mono Lake and other lakes when the water was fresher and to have adapted as the salt concentration rose.

Most brine shrimp belong to the species *Artemia franciscana* and disperse by air, with dormant embryos in small cysts that resemble a seed. The wind carries them everywhere, and the lucky few that land at salt lakes hatch and survive. But Mono Lake's shrimp are a slightly different type than those of other lakes. They're adapted to live in Mono's special chemistry, which is more alkaline than the chlorine-dominated waters of other lakes. They have their own species—or subspecies, the scientific debate is still going on—called *Artemia monica*, which was first identified by Dana in her master's thesis. (She also showed that algae abundance is the limiting factor for brine shrimp populations, which explained why the concentrations of shrimp at Mono were so much higher than those at Great Salt Lake.) *Artemia monica* cysts don't disperse; they stay at the lake. Adult shrimp lay them as the weather gets cold in the fall, before dying off. The cysts stay dormant until the spring, when the yearly cycle starts again.

The shrimp thrived at the lake's salt concentration of 12 percent. But the student researchers estimated that, at its current rate of decline, Mono's salinity was going to quadruple. To understand what effect the increasing salinity would have on the creatures, Herbst and Dana went to the chemistry lab at the high school in the nearby town of Lee Vining with canisters of Mono Lake water, which they boiled down into samples with a variety of salt concentrations. The consequences were dramatic. When they doubled the salt concentration, the shrimp died off en masse as the concentration of salt in their bodies multiplied twenty-fold. They'd adapted to regulate the salt in their systems up to a certain threshold; above that level, they began to shut down. "The present populations of these animals will not be able to withstand the increasing salinity predicted for Mono Lake," Herbst and Dana wrote.

The situation was similar with the alkali flies, the lake's only other invertebrates. There, Herbst took charge, earning him the lifelong nickname Bug. (The other David in the group, David Winkler, became Wink.) Where brine shrimp are among the earth's oldest creatures, flies are among the most recently evolved. Flies arrived at the lakes once they were already saline, and evolved into a new species that could survive there.

When Bug set out to understand the flies, he learned that their life cycle is bizarre. They lay their eggs on the surface of the lakes. When the larvae hatch, they swim in the water and crawl around the tufa towers and the rocks of the lake's shallow shore. The larvae can breathe underwater: their respiratory tube evolved over time into a gill at the trachea. They attach themselves to a solid site using their rearmost pair of prolegs—fleshy, foot-like organs similar to the feet of a caterpillar—which are larger than the others and have special claws that can grasp the rough tufa. Then, they build a cocoon and enter their pupal stage. The pupa dissolves into "biologic dew," a bubble made of amino acids and a few cells with instructions on how to become a fly. A balloon-like structure pops out from a plate on their head, the head gets put back together, and the fly is born in the air bubble. It floats to the surface to live as an adult for a few days, returning to the water to lay eggs or feed on algae thanks to tiny hairs that trap water and form a hydrodynamic surface around it that a news segment called a "shimmering bubble shield." Within a matter of days, it becomes food for birds.

Bug dissected the flies to understand how they survived in the salt lake. He discovered that they have a specialized kidney with tiny tubules that filter out the salts. When he repeated the shrimp experiment with fly larvae, introducing them to buckets of increasingly saline water, he discovered that, like the shrimp, the flies were happiest at 12 percent salinity. Above that level, they began to go dormant and stopped pupating. As the lake got saltier, then, both the shrimp and fly populations would decline. That would mean a major blow to bird life—the aspect of the food chain for which Mono Lake is most famous.

Over the course of the summer, the students carried out five all-lake bird censuses. They used those counts to develop new estimates of the maximum daily populations of the bird species that used the lake, tallying 46,000 gulls, 93,000 Wilson's phalaropes, and 750,000 eared grebes. The numbers were unexpected. The gull count alone was ten times higher than past estimates; it made Mono the world's second-largest California gull colony, representing 85 percent of the state's breeding population. They also added snowy plovers to the list of birds that bred at the lake.

In addition to counting the birds, the students captured individuals with nets to measure their body condition and fat content. Then, they took diet samples by putting straws down their throats and extracting the contents of their stomachs. They observed that gulls ate almost exclusively shrimp, while phalaropes opted for flies. (Later, scientist Margaret Rubega discovered that this is because the flies are more nutritionally dense: the tiny birds can't fit enough shrimp in their intestine to get all the calories they need.) Based on the phalaropes' weight and the length of their wings, they calculated that they could make it from Mono to the Salton Sea or the Gulf of California without needing to touch down.

The students were never able to catch an eared grebe, which were more numerous at the lake than any other species—and the species that would be the most impacted if the lake were to disappear. The small black waterbirds, with red eyes and a distinctive ocher-colored fan of feathers that resemble an ear, are among the most halophilic, or salt-loving, species in the world. Their entire population depends on just a few saline ecosystems. In late fall, around a third of the world's eared grebes camp out at Mono Lake, molting and feeding on shrimp until they double their weight. In general, it's rare for more than 1 percent of any bird species' population to be found in once place at one time. But since there are so few salt lakes, the species that depend on them only have a few places to go. If Mono were to become too saline for flies and shrimp, the students realized, 30 percent of the world's grebes could be lost in one blow.

THE MORE I LEARNED about the creatures at Mono and their adaptations to the lake's strange, salty conditions, the more they reminded me of a fringe body of theory I'd read about called queer ecology. It spanned being a perspective on science, a per-

spective on the world, and a mode of being in the world. As a
perspective on biology, it directed attention to the myriad creatures of the world that don't conform to "normal," heterosexual
practices of sex and reproduction. All of the salt lake creatures
seemed to be queer in this way—the freakish denizens of "freakish habitats," as Rubega had put it.

Brine shrimp have perhaps the most manifold reproductive systems of any creature I'd heard of. Some are bisexual, meaning that
they have both male and female sex organs. Some are partheno-
genic, meaning that they reproduce asexually, without fertilization.
They can also birth their young in two ways: the cysts they lay in the
fall to overwinter and hatch later are oviparous, meaning that they
hatch outside the parent's body, but during the summer, the adults
that have hatched from cysts birth ovoviviparous young—born as
live nauplii, instead of as cysts. The creatures have subverted the
rules of reproduction, finding every possible route and adjusting
their plans depending on the conditions of their environment.

Phalaropes, too, are unusual; they have reversed gender roles.
Though ordinarily male birds are larger and showier, among phalaropes it's the females who have the bigger bodies and bright colors. The females also practice polyandry, or the taking of multiple
male partners. After laying their eggs they abandon the nest, leaving the males to brood and take care of the young.

These adaptations—"biological exuberance," as biologist
Bruce Bagemihl titled his book about homosexuality in the natural world—resonate with ideas of queerness not only because
they mirror humans' diverse sex practices, family structures, and
reproductive adaptations, but also because they enable creatures
to adapt to adverse conditions, like salt lakes' saline water, low
oxygen, high ultraviolet radiation, and extreme temperatures. The
queer creatures that have managed to survive at the lakes have
done so by being creative: the flies excreting salts from their tiny

insect kidneys, the shrimp synthesizing the hemoglobin in their blood in order to store oxygen and release it slowly. Not only that, but they have survived in abundance. They have flourished, proving that places that appear sterile can be full of life.

But just because they have found ways to tolerate saline ecosystems doesn't mean that they can survive anything. The lake's creatures are sensitive to changes in their habitat, and the salinity was fast becoming more than they could handle. If the salinity kept increasing, the young scientists worried, there would soon be no shrimp, no flies, no grebes or phalaropes or gulls.

They needed to convey this to the public. In the fall, Wink collected all of their findings into a comprehensive document. David Gaines, the ornithology master's student who had spearheaded the grant application, convinced the Institute of Ecology at UC–Davis to publish it. But the group's findings wouldn't help save the lake until things got worse.

THE FOLLOWING SUMMER, Wink and his roommate Don made a visit to Mono. The water level had dropped even further. The real shock came when they realized they could walk out to Negit Island on a land bridge. Negit was the lake's major nesting site for California gulls. Its rocky surface was blanketed white with guano and its skies were full of the white birds with black-tipped wings. What made it such an important site was that it was an island, and therefore was protected by water from coyotes and other predators. With the water levels so low, if they didn't do something, the gull colony would collapse.

"We gotta go get Gaines," Wink told Don. Gaines had holed up in Northern California's Eel River Preserve, seeking to get as far away from civilization as possible. There was no way to reach him besides getting in the car and driving nine hours. So they did. "And there, in the caretaker's kitchen, we basically founded the Mono Lake Committee," Wink recalled.

They developed a strategy that would combine publicity, visits, research, lobbying the state legislature over air pollution concerns, and finding a basis on which to take the Los Angeles Department of Water and Power to court. The first order of business, though, was to design bumper stickers. Wink objected to SAVE MONO LAKE because, he told me, if they could just maintain the status quo, it would be okay. They settled instead on LONG LIVE MONO LAKE—the bumper sticker I'd seen on the back of Dylan's friends' car. "It was an expression of our love for it," said Wink. "But it had to become 'Save Mono Lake' when shit hit the fan."

Wink, Herbst, and Dana all ended up going to graduate school—Wink to study gulls, Herbst to study flies, and Dana brine shrimp. "Gaines was sort of left holding the bag," Wink told me. The following year, he and his future wife, Sally, bought a pair of houses in Lee Vining and set up a headquarters from which to run the Mono Lake Committee. He went on the road for eighteen months, narrating a slideshow about Mono Lake at schools, colleges, and chapters of the Sierra Club, Audubon Society, and Lions Club across California—but especially in Sacramento, so that lawmakers would get wind.

As the committee took shape, its members joined forces with Such and the others working on a public trust case. Finally, on May 12, 1979, lawyers from Friends of the Earth, the National Audubon Society, and the Mono Lake Committee went to the Mono County courthouse to file suit against the Los Angeles Department of Water and Power. Just a month earlier, workers from the state fish and game department had found Negit strewn with empty gull nests and broken shells, and traversed by coyote tracks.

The coalition of environmental organizations argued that Los Angeles's rerouting of water headed for Mono Lake had violated the public trust doctrine back in 1940, when the agency received its diversion licenses from the state. Public trust trumped prior appropriation, they argued; therefore, the state water board had the power to stop the water exports.

LADWP argued the reverse: the city agency claimed its water rights superseded the public trust doctrine. California water law said that water had to be put to beneficial use; under the "use it or lose it" interpretation of that concept, cutting back their use and letting some water flow into Mono would legally be considered "waste" and they could lose their water right. Public opinion, they claimed, supported this: Mono was just a smelly, worthless salt lake that wasted useful water; serving people was its highest beneficial use. Besides, Eastern Sierra water was much cheaper for their customers than any other alternative; other sources might not even be available.

The case went to the California Supreme Court as *National Audubon Society v. Superior Court*. In February 1983, the court made its decision. The justices wrote that the public trust and prior appropriation were on a "collision course" in state law, and that determining which one had priority was an urgent matter. They chose the public trust, and ruled that the state had the obligation to "avoid or minimize any harm" to public trust values, even if it had to re-appropriate water rights to do so—not only at Mono, but everywhere water was allocated. They affirmed in-stream flows—the legal term for water that stayed in the river and flowed all the way to the lake—as a beneficial use, and determined that privately owned water rights were not permanent and inalienable but subject to state supervision and reconsideration.

The court left the specific terms of Mono's management and the re-appropriation of Los Angeles's water rights to the California State Water Resources Control Board. For years until the board came to a decision, the court's ruling had no impact on the ground. Los Angeles kept diverting water, and Mono's water level kept dropping. Finally, in 1994, after state- and federally-funded scientific studies confirming the imminent death of the lake's ecosystem, threats of legal action over the board's sluggishness, and separate lawsuits over fisheries in two of the waterways that fed Mono, the board made its announcement. It opted not to take back Los Angeles's senior water right, but obligated the agency to cut its diversions by two-thirds. It also set a target management elevation for the lake of 6,392 feet above sea level.

The decision enshrined the public trust doctrine as a concept with the legal teeth to protect salt lakes and other waterways. It made water once again a matter of the common good, rather than private property. "Every single water license issued now has a clause about the continuing jurisdiction of the state water board over water rights and the capacity to change them," Geoff McQuilkin, current director of the Mono Lake Committee, told me. "Before, no one was going to mess with you until the end of time."

GAINES DIDN'T LIVE to see the final determination. In 1988, he and a Mono Lake Committee staff member died in a head-on collision while driving on Highway 395 during a snowstorm. His wife and two children survived the accident. He had devoted his short life to the Mono Lake Committee. "And he did an amazing job—especially amazing when you realize that he wasn't well suited for the incredible amount of people-ness," recalled Wink. "He sacrificed even more than it would appear, because it was a mismatch for his personality."

Today, the Mono Lake Committee operates an information center and environmental bookstore out of a Mission-style storefront in Lee Vining. Their staff of scientists, educational program leaders, policy specialists, and administrators swells to thirty in the summer season. They sponsor a birding festival, an annual run, scholarships, and awards. But in the end, Dylan and I didn't stop there. We ate the famous fish tacos at the nearby Mobil station and drove through Yosemite in the dark, pulling into Merced at quarter to midnight. It would be years before I visited the lake. But even catching it out the window astonished me: the purple, green, and yellow landscape and the reflective blue water, the crater-like bowl and the vivid alkali grasses etched themselves into my mind's eye. There was no question as to why they'd fought to save it.

CHAPTER 5

RESTITUTION

Zuni Salt Lake, New Mexico, USA

AT THE SAME TIME AS THE MONO LAKE COMMITTEE was coming together to sue Los Angeles, a very different group of activists was fighting a very different battle for a salt lake.

The Zuni people originated in an underworld near the Grand Canyon. Over time, they migrated through Arizona, New Mexico, and Colorado to the area of western New Mexico now known as Zuni Pueblo. "We were led to the place which we know to be the center of the earth," said Mekalita Wastaluci in 1977. "The rainbow directed us there."

Just south of Zuni Pueblo is a salt lake. In Zuni belief, the lake is Ma'l Oyattsik'i, the mother of the Zuni people. Historically, members of the tribe made a yearly pilgrimage to the lake. They reached it through a vast network of trails spanning about six hundred miles, marked by cairns that were carefully constructed, with prayers accompanying each added material. The pilgrimage was made in early summer, shortly after the planting season— visiting any other time was sacrilegious—and was undertaken only by members of the Rain Priesthood. Women were not allowed, and are still not allowed to this day. When the men set off, the entire pueblo filed into the streets and onto rooftops to watch their departure and offer them prayers. They reached the lake on the second day of their journey and spent the night with Ma'l Oyattsik'i, who comes out after dark to visit the supernatural beings who live at nearby Zuni Heaven and to accept the offerings left for her.

By the early twentieth century, however, Spanish, Mexican, and US colonialism had taken the lake away from the tribe and impeded Zuni access to the sacred site. In the 1970s, as scientists and environmentalists eight hundred miles away wrestled with the public trust doctrine, the tribe set out to get it back.

I ENDED UP in New Mexico by a stroke of luck. After spending the summer at Dylan's parents' house, in early September we drove back to Denver. He would fly from there to England to finish his master's degree; I didn't have any plans.

Days before he left, we went to the wedding of two of his friends, a couple who jointly managed a cattle ranch. By chance, the ceremony was held at the home of the bride's grandmother, several miles south of my dad's house. Soon, we were sitting at a folding table under a cottonwood tree, eating Mexican food from the wedding buffet with Dylan's college friends, all well-read nerds with a sense of humor and a style of dress similar to his.

Nick, the groom, sat down at the table with us. He was six feet tall and had exactly the kind of lanky build I pictured cowboys having.

"It turns out weddings aren't a great place to see your friends," he said, apologizing for the fact that everyone had traveled there to be just one stop in his circuit of tables. One of Dylan's friends asked him how ranching was going.

"We're actually looking for a new ranch hand, if any of you wants to come down," he said, playing jokingly off their interest. "Ours is leaving in two weeks."

"I could do it," I blurted out.

I'd dreamed of ranching long before I understood what it was. When I was nine, my mom had sent me to summer camp in the mountains, largely to have the experience of riding horses. It had turned out to be an early experience of class consciousness: I remember restraining my full-bodied enthusiasm at getting to ride, not wanting to be mistaken for one of the camp's more spoiled, horse-obsessed girls. Instead, I code-switched and befriended the misfits: one daughter of a small-town postal worker and another from the exurbs of Denver whose father had died that year. But I also felt a disappointment when we did ride horses. We were always going around a ring or following a leader in a line along forest trails. In my mind's eye, riding entailed galloping through an empty field of tall prairie grasses, simultaneously on a mission and completely free.

Where had I come up with such an image? Perhaps from splicing together the cowboy imagery ever-present in my Western childhood with my family's nostalgia: the red copy of *My Friend Flicka* my grandmother gave me, hers from her own Denver childhood; our yearly excursions to the National Western Stock Show; my grandfather's stained 35mm films of herding sheep on his family's ranch in Wyoming's Wind River Valley; my parents' sadness about the suburban sprawl surrounding Denver, their memories of ranches so close by giving a feeling of sacred openness to Colorado.

The vision stuck with me. Then, as a high school student, I read the book *The Last Ranch*, about a southern Colorado rancher's attempts to use unconventional grazing methods to restore the tallgrass prairie. The next year, I found *The American West as Living Space*. The West has "boomers" and "stickers," Stegner writes: those who come looking to strike it rich and then move on, and those who stay and learn to live with aridity. To him, ranchers are the latter—those who have learned to live with the landscape. "Ranching is one of the few western occupations that have been renewable and have produced a continuing way of life," he writes.

I wanted to be a cowboy because I wanted to be a sticker. I wanted to be a cowboy because I thought it was the most authentic and challenging thing I could do. I wanted to be a cowboy because I wanted to earn the West like a merit badge. But I'd never known how to find a way in.

"For real?" he said.

"Yeah," I said. I tried to act as though he weren't offering me what had been my dream job since I was eight. We exchanged emails and agreed that I'd start a week later.

From my first day, my body tired from the inside out, I felt as though I'd come to the precise place I had always dreamed of. The blue skies, the expansive grasslands, the uncharismatic town, the animals—I felt a sense of belonging that penetrated deep into my heart.

The ranch spanned 13,000 acres of northeastern New Mexico's shortgrass plains, east of the mountains in a town where two highways crossed. It had none of the Southwest-themed tourist appeal of Santa Fe or Taos: even the #NewMexicoTrue billboards flanking the highway featured images of Shiprock and Gallup, on the other side of the state. Headquarters were a short drive east of town, a sea of blue grama, alkali sacaton, ring muhly, four-wing saltbush, winterfat, and other grasses and forbs. Amy, Nick's wife, said she'd never seen such a diversity of grasses on one ranch. It was a beautiful landscape, but one of the first things I noticed was that no one used that word. They only ever called it "country," like they were accidentally, organically bringing to life the dictum of Welsh cultural critic Raymond Williams that "a working country is hardly ever a landscape."

During the summer, the twenty-four bulls and 330 cows grazed an enormous mesa that was a forty-five minute drive away and had a distinct ecology: rugged slopes of scrub oak forest and large, treeless draws on the sides and a flat grassland tabletop with cooler weather, fog, and theatrical lightning storms.

Amy and Nick were both first-generation ranchers, and after years of apprenticeships, it was their first time managing a ranch on their own. In addition to us three humans, the horses we rode, and the 300-odd head of cattle, there were three cattle dogs: Liz, Puck, and Bat Dog. They worked alongside us, ranch hands in their own right, with a special set of commands: "come by" to move clockwise around the herd, "away to me" to run counter-clockwise, "walk up" to approach the cows slowly, "get out" to bring in a straggler, and "that'll do" to indicate their work was done. They coordinated between the humans, the horses, and the cattle, like one multi-species organism accomplishing a single task. Liz, short for Lizard, was the alpha of the group—the smartest and most experienced—and she knew it. Puck, barely more than a puppy, was always "bein' brave," in Nick's words— enthusiastic to the point of recklessly showing off. Bat Dog, who belonged to Amy, was quiet and kind in the same way as her human counterpart. One day when I was sick in bed, she appeared in my room to check on me. She looked at me for per-mission to jump onto the bed, which I gave, and she kept me company. Some months after I left, Amy wrote to me to say that Bat had slid off the back of the truck while she and Nick were helping at another ranch on a muddy day, and had been run over and killed. It was the first time I'd mourned an animal the way I would a human.

MY SECOND WEEK on the ranch, it was time to bring the cows down from the mesa. That meant a cattle drive: eleven miles along the county road to one of the back gates. Cowboys from other ranches were coming to help, and the sheriff was going to block traffic when we had to cross the highway.

As we funneled the animals out of the pasture gate in single file onto the road, the others began teaching me how to move cattle. They told me to make the sound *sh, sh, sh* behind the animals to keep them moving, to ride back and forth along the segment I'd been assigned to keep them in formation, and to bring those that wandered off back into the line by circling wide and then riding counter to the herd's current, passing through their lateral field of vision.

The drive lasted hours. As the sun came out, we peeled off our gloves and tied down our parkas behind us with saddle strings. By the end, I felt exhilarated. Everyone did. New to the work, I didn't realize until later that it was monumental for the others, too. Most ranches that have higher-altitude summer country ship their animals back and forth in semi-trailers; highways and fence lines have made long cattle drives too challenging. Even the seasoned hands felt like they had realized the fantasy that had led us all to ranching.

Moving cattle never stopped being exhilarating, even when the distances were shorter. Each week, we moved the cows two or three times between pastures within the ranch. If the pastures were next to one another, we could simply open the gate and call them through, yelling "WAY-OHHHHH!" again and again, throwing the air out of our lungs in a complete and shameless release. When we had to cross longer sections of the ranch, the moves required more maneuvering on horseback to gather and drive the animals, with one of us at the head opening the gates and the others at the back keeping the forward motion and sending the dogs to round up stragglers. It became my favorite feeling in the world: the way the thousand hooves patted the ground with odd softness,

the way the cows moved away from the horses like magnets repel-
ling one other, the way nuanced movements on horseback could
adjust the shape of the herd.

After the cattle drive from the mesa, the next big event
of the year was preg-checking, in November. The owner of the
cows, Nick and Amy's boss, came to do the honors—sticking a
gloved arm up the anus of the cows—himself. He called them
names as he did: "fatso," "sister," "bred red," "beauty queen."
With the results, we separated the cows into two herds, the pregnant cows and the "opens"—those who had failed to get pregnant
during their summer sharing a pasture with the bulls. In January,
we divided the pregnant cows again, into first-time mothers and
experienced ones.

In March, calving season started. In addition to all the regular
tasks, we had to ride through the herd twice a day, at dawn and
dusk, to check for cows in labor and make sure nothing was going
wrong. Before the season started, I wondered what we would stop
doing in order to make the time for this extra work. The answer
was nothing; we just took on more than we could handle. I had
never been so overwhelmed; I often wanted to cry all day from the
stress. One day, I did cry the entire day, thicker and heavier and
more uncontrollably than I had ever cried, as Nick and I drove
around doing chores; he said nothing, for which I was grateful.

Normal births were over, unassisted, in half an hour; the
mother would strain and the calf would drop out headfirst, its
tiny hooves pointed into a swan dive. The mother would lick the
birth sac off its hide and within minutes, the calf would stand
up and take its first drink of colostrum. But there was a lot that
could go wrong.

One calf was born premature; we put her in a dog bed in the

mudroom of my house and bottle-fed her for days, until one day we went into town and came home to find her dead. Multiple mothers refused to nurse their young. After giving their calves formula colostrum that we mixed in my kitchen, we would immobilize the mother in the chute, the narrow metal contraption used to queue and restrain cows for procedures like pregnancy checking and vaccination, and open a flap of metal on the bottom of its left side to allow her calf to "get a drink," as Amy put it.

Other calves needed to be pulled. Large bulls belonging to a neighbor on the mesa had broken into our pastures during the summer and bred a number of the cows, who gestated calves so large that they struggled to birth them. One night after an emergency call from Amy, I drove out and helped pull a hundred-pound calf that we named Hulk. The hard labor had given the mother temporary nerve damage, meaning that she couldn't stand up. So, every day for two weeks, we drove the tractor to her, tightened a metal frame over her hipbones, and lifted her to a standing position for thirty minutes, allowing her to start to move her legs again. Finally, one day, we arrived to find her already standing. Meanwhile, we cared for Hulk at headquarters.

Days later, a pleasant spring evening bore a surprise snowstorm overnight. The ranch was beautiful in the snow, like a completely different place. That morning, riding in our ankle-length slickers, we found that a cow had birthed a stillborn calf but tried to nurture it anyway, licking and licking its dead hide. We tied a rope around the dead calf and dragged it behind us over the white ground, the mother following us to the corrals. Back at headquarters, we skinned the dead calf and stretched its wet hide onto Hulk. We led him in his little vest to the dead calf's mother. It worked: she smelled him as her own calf and let him nurse. Within a few days we pulled off the hide and released them back into the herd as a pair.

Another night, I got back from checking my half of the cows, took a shower, and made enchiladas for dinner. As I was about to pull them out of the oven, Amy called me. She was at the corrals. When she was out checking the cows, she'd seen a heifer that she could tell had been in labor for hours; a stillborn calf was hanging halfway out of her vagina. My clean hair in a wet braid, I rushed over. She'd penned the mother on horseback and was standing on a rung of one of the metal fencing panels, trying to loop a rope around the mother cow's neck so that she could get her to lie down and we could pull the calf. But the mother kept charging at the panel, forcing Amy to jump away.

We got on horses and guided the heifer into the chute. She fell down. Amy sighed, exasperated. Finally, I got the calving chains from the barn and we looped them around the calf's dead metacarpuses. Each taking one hook, we braced our feet against the chute for leverage, and pulled out the seventy pounds of dead weight.

Then things got worse. The mother began to prolapse: she had strained so much that she had pushed her uterus out along with the dead calf. It was 9 p.m. and we had lost our light. Amy stayed with the mother while I ran for lanterns, iodine, rags, and a plastic tub partially filled with water. When I got back, Amy was holding the uterus, which was slowly spilling out of the mother cow, off the ground. It was covered in oval-shaped purple lesions. She had some kind of bovine Kaposi sarcoma, I thought, recoiling.

"They're cotyledons!" Amy said, in a tone that suggested I needed to get over myself. We needed to pull the placenta off of the uterus, she explained quickly. The cotyledons were normal. In humans, the placenta grows inside the uterus, but in cows, it grows alongside it, connected by the cotyledons and their inverse, caruncles, which snap together to allow for the transfer of fetal blood, oxygen, and nutrients. Separating them felt like pulling apart some kind of slimy alien Velcro.

As the uterus slowly fell out, we caught it in the plastic tub and splashed it with iodine and water. It was round, firm with a soft exterior, and weighed as much as the calf—seventy pounds. It reminded me of a medicine ball, and as we started putting it back into the vagina, I used the same technique I use to push my sleeping bag into its stuff sack. For every piece we pushed in, another spilled out.

Finally, we got it in. Amy stitched up the vagina. We stood up and looked at one another. We were covered in blood. It was 11:45.

"See you at 6:30," Amy said, exhausted and nauseated.

As TIME PASSED, the way time passes on ranches—cycling slowly at the scale of the seasons and repetitively at the scale of the week—what came to matter to me was not living out the cowboy fantasy I'd come looking for but the work's quotidian satisfactions: the sensory saturation, the relationships with the animals, the fresh air, the plants. Ranching required dense, tactile forms of thinking that were new to me: animal husbandry, grassland ecology, irrigation lines, accounting, truck mechanics. Hundreds of beings depended on us to do even rote tasks correctly. I'd never had a job where the stakes were life and death. The threshold for frustration was higher than it had ever been in my life. I learned to stay calm, because yelling made dangerous situations worse. I learned to dress for the weather, to slow down and be thorough, to solve kinetic problems, none of which came easily to me. I learned to fix fence in a way that would keep it intact for years and to find shorts in the electric line, which felt like solving for x in algebra. I checked the trucks' tire pressure every morning, even when I didn't feel like it.

I never felt lonely. I was so immersed in this new world that I didn't mind growing more distant from Dylan. I spent my days off in a blissful solitude, snuggled in front of the wood stove, reading. Without telling anyone, I'd begun downloading the classics of queer theory on my iPad. I took handwritten notes in the journal I used to log my days as a cowboy. After spending my days outdoors, working with my body, nothing was more pleasurable than trying to wrap my mind around the convoluted sentences of critics like Leo Bersani and Hélène Cixous. Whereas in high school and college queerness had felt like a community to which I couldn't belong, alone on the high plains it felt like the only way to live. Suiting up in men's Wranglers every day, off to shed a learned helplessness I'd never realized I had acquired, allowed me to inhabit myself queerly before making any utterance to the effect.

Most of all, though, the work changed my relationship to the landscape. It humbled me within my ecosystem. From the first week of work, I had to begin memorizing the minute distinctions between the ranch's grasses and forbs, learning whether they were warm- or cool-season, how hardy they were and whether the cows liked them. With Nick and Amy, I had to make decisions at the scale of an ecosystem: would we defy the landowner's instructions to demolish the beaver dam in order to keep water flowing to the irrigated hay field? "I kinda like 'em," Nick said, even though most ranchers saw beavers as problems. "They were here first, right?"

I watched how the seasons changed, and how the animals reacted to those changes. In our relationships with cows, dogs, grasses, pronghorn, rivers, soil, and many other more-than-human beings, ranching felt like what feminist theorist Donna Haraway calls "significant otherness": "vulnerable, on-the-ground work that cobbles together non-harmonious agencies and ways of living that are accountable both to their disparate inherited histories and to their barely possible but absolutely necessary joint futures."

THOUGH I FELT MYSELF "becoming with" the landscape—again in Haraway's words—I also began to feel more than I ever had like the settler colonist that all of us non-Indigenous Westerners are. At the same time that moving cattle on horseback felt like a delicate and satisfying dance, it also taught me, in a bodily way, what it meant to dominate a landscape. The contradiction between the poverty and racialization of the nearby town and the massive, Anglo-owned ranches surrounding it began to forge a loud cleavage in me.

By the spring, I had traded queer theory for the beginnings of an academic bibliography on New Mexico, starting with David Correia's *Properties of Violence* and Angela Garcia's *The Pastoral Clinic* and continuing to Maria Montoya's *Translating Property*, which told the story of the very land grant where the ranch was located, on which I was spending my days moving cows, sorting scrap metal, and adjusting the irrigation headgate—its transformation "from an 'empty' landscape that housed the semi-settled Jicarilla Apache to a hacienda run by Lucien B. Maxwell and finally to a land company run by Dutch overseers."

Then I learned that, on the other side of New Mexico, there was a site where the theoretical questions I was wrestling with from my cozy, green-carpeted room had played out in actuality.

THE ZUNI SALT LAKE is a maar, or a broad, shallow volcanic crater filled by a lake. It is about a mile in diameter and just over two feet deep, and sits at an elevation of 6,200 feet. The steep walls around it stand some 150 feet high. The walls and crater floor are made of sandstone covered by malpais, or dried black lava. Piñon, juniper, saltbush, and grama grass fill the landscape. According to geologists, the maar formed 23,000 years ago through phreatomagmatic eruptions, which geologist Spencer Lucas explained to me take place when magma beneath the earth's surface comes into contact with the water table and boils the groundwater until its steam explodes through the earth's crust. When the ground cools again, it settles into a crater.

New Mexico has about twenty such maar craters, but most of their lakes evaporated at the end of the Pleistocene. Back then, Zuni Salt Lake was much larger—about three times its current size. The lake continues to be fed by surface runoff and mineral springs, which Lucas calls "groundwater leaking." Its waters are 99 percent sodium chloride, or table salt, that comes from the bedrock beneath the crater. Its waters support brine shrimp, brine flies, bacteria, and algae. Two cinder cones rise out of the lake; one has a crater holding another pool of saline water that is warm and red because of blooming sulfur bacteria.

Zuni Pueblo's ancestral territory surrounds the lake. On its western edge, it extends from Arizona's Gila River north to the tribe's site of origin near Tuba City; to the east, it reaches as far as New Mexico's Sandia Mountains. It's a landscape where the Great Basin and Colorado Plateau ecosystems merge, a landscape of *Artemesia* (sagebrush) and *Atriplex* (saltbush), rabbitbrush, winterfat, Indian paintbrush, mountain mahogany, wolfberry, prickly pear, chokecherry, oak, greasewood, willow, and yucca. Many of its grasses are the same as those of the ranch where I worked, all of which I came to love: Western wheatgrass, little and big bluestem, black and blue grama, sideoats grama, galleta grass, ring muhly, ricegrass, alkali sacaton, sand dropseed, needle-and-thread grass, New Mexico feathergrass. There are mule deer, black bears, coyotes, porcupines, foxes, rabbits, and many types of mice. There are hawks, eagles, owls, violet-green swallows, rock wrens, roadrunners, flickers, flycatchers, and vireos.

In Zuni belief, Ma'l Oyattsik'i, the salt mother, once lived in a different location. She moved because humans disrespected her sacred skin. But she left a trail, called Malo Keseke Onnane or Salt Woman Trail, that led across the landscape to her new home. "The people who lost her began searching," a Zuni cacique, or religious leader, named Smith Cachane explained in 1977. "They made their prayersticks and the Head Priest and another started out riding donkeys. They traveled all day, and the following day they came upon her. There they planted their prayersticks, praying for forgiveness for being so careless about keeping the lake clean and beautiful."

Historically, not only Zuni, but Acoma, Hopi, Laguna, and Taos Puebloans, as well as Navajos and Apaches, made annual pilgrimages to the lake, traveling hundreds of miles to harvest the flakes of dried salt that for the Zuni constituted the flesh of Ma'l Oyattsik'i. (The name Zuni was given to the tribe by the Spanish, derived from a mishearing of the Keres Pueblo's name for the tribe; the tribe's name for itself is A:shiwi, which means "the flesh.") Each culture had its own beliefs, ceremonies, and rituals associated with salt gathering. They left offerings of prayer plumes, shields, and bows and arrows along the margin of the lake. It was a rare neutral zone where hostilities between tribes were suspended. Apache chief Geronimo wrote in his autobiography, *Geronimo's Story of His Life*: "When visiting this lake our people were not allowed to even kill game or attack an enemy. All creatures were free to go and come without molestation."

Once they arrived, the men performed ceremonies and made their offerings. The Zuni men planted prayer sticks on the altar at the edge of the lake, and then stripped naked to enter the lake and harvest the salts. They had to be respectful of Ma'l Oyattsik'i, who had already fled humans once. "As a race of Zuni people we don't intend to be destructive to our mother," tribal member Alex Seowtewa told historian Richard Hart in 1980. "Even a little tree has to be taken care of by giving it cornmeal with a mixture of turquoise before being used, as an offering to the spirits of our ancestors and to the spirits of Gods." Those who were visiting for the first time lay face down and struck the water six times before entering. The cinder cones were considered by the Zuni to be the homes of their war gods, and the small salt pond inside one of them was even more sacred than the large salt lake: only a few high priests were allowed to descend into it.

The Zuni collected the salt by hand, washed it in the lake using sieves made of yucca, and packed it in blankets to transport it on the backs of donkeys. For the Acoma visitors, the harvest was solemn: they were not allowed to laugh as they worked. If the lake was too deep to harvest, the Hopi visitors held additional ceremonies to ask that it would recede. And as they left, they were not permitted to look back at the salt because if they did, their spirit would remain trapped at the lake.

In 1540, the Puebloans' world started to change. Spanish explorers passed through Zuni Pueblo and followed the salt trails to the lake in their search for the Seven Cities of Cibola, where they believed they would find gold. The conquistador Francisco Vázquez de Coronado praised the salt from Zuni Salt Lake as "the best and whitest I have ever seen in all my life." In 1598, New Mexico's first governor, Juan de Oñate, heard about the salt lake while in Zuni demanding vassalage, and sent a member of his party to find the salt. He returned with the same impression: that it was a massive, marvelously white deposit.

Though the Spanish government nominally granted the Zuni rights to all the land they used and occupied, colonial rule nevertheless meant encroachment onto both the tribe's territories and belief system. The first Catholic missionaries arrived in 1629 and set up two churches. In 1632, the Zuni killed the friar installed to convert them. Afterward, fearing retaliation, they fled their villages and lived for three years atop Dowa Yalanne, or Corn Mountain, a mesa three miles southeast of Zuni Pueblo. When they returned to the bottomlands, they faced increasing hostility from the Apaches, themselves driven to violence by encroachment by the Spanish. The Zuni decided to ally themselves with the Spanish Empire for protection. But in October 1640, Apaches raided the Zuni village of Hawikuh and killed the new friar, beating his head with a bell while he clung to a cross. The Spanish temporarily abandoned the Zuni mission. When they installed a new priest, the Zunis murdered him, burned the church, and fled again to Corn Mountain. They stayed there for twelve years, until New Mexico governor Diego de Vargas climbed the mesa in 1692 to baptize the Zunis and demand their loyalty. They returned to the flatlands but lived only in their three easternmost towns, fearful of becoming too dispersed and therefore unable to defend themselves against the Spanish or the Apaches. Gradually, they consolidated into a single town.

In the early 1700s, the cycle repeated. The Zunis' new priest wrote to Santa Fe that he was concerned about Apache depredations. The colonial government sent three men to protect him, but Zunis intercepted and killed them on their way in. Afterward, the tribe fled once again to Dowa Yalanne. The friar convinced them to return, but they continued to live in a state of fear. Despite the vastness of their homeland, they remained huddled in a single village, with men leaving in the summers to tend their fields and sheep.

Throughout this period, they continued to visit the salt lake and hold on to their beliefs, even as the Spanish government and its priests destroyed their shrines, structures five feet high and two feet thick made from the lakebed's blue clay. Their devotion to their traditions pushed the Catholic priests to desperation. In

     SALT LAKES

1805, Fray José de la Prada wrote that if being a missionary at Zuni "had been chosen [as] a prison for those guilty of the gravest crimes, there would not have been a more severe decision." He added that "in the midst of such distress, they have the greater one of each day seeing the Zuni Indians become more backward in the divine Christian religion, due to the repugnance that they show for the divine law."

In 1846, the United States invaded Mexico to claim what is now the American Southwest. The Treaty of Guadalupe Hidalgo, which ended the war two years later, required that the US government would recognize all Pueblo Indian lands to the same extent they had been respected by Spain and Mexico. The government established the post of surveyor general of New Mexico, one of whose duties was to investigate Pueblo land claims and report them to Congress for confirmation. The Zuni submitted papers substantiating their claim to the surveyor's office, but in the end he reviewed all the claims in New Mexico except theirs, because he was worried about violence in the region by Navajos and Apaches. The fears weren't unfounded. When the US Army visited the area—including the salt lake—in 1856, the government's agent to the Navajos was killed by Apaches.

By the early 1870s, the Commissioner of Indian Affairs had visited twice and reported that there were 1,530 people in Zuni Pueblo, 333 of them children, and that they had never seen another government agent in the village. He wrote, "They have occupied the country in which they now live for more than two hundred years, and they earnestly request that Congress grant them a reservation and give them a title to it." But when the Wheeler Geological Survey passed through the region in 1873, the surveyor recommended only a small reservation—2 percent of the tribe's homeland—without ever consulting with the tribe. The reservation was established by President Rutherford B. Hayes in 1877, depriving the Zuni of some 9 million acres of ancestral lands.

The tribe complained to the state surveyor general's office, concerned particularly about the fact that the reservation didn't include the salt lake. Their concerns fell on uncooperative ears. The surveyor general was known to have been involved in the territory's rampant land frauds. In 1879, he certified a land grant as genuine that even his contemporaries knew was a forgery; meanwhile, all the Spanish-era documents relating to the Zuni—including a real land grant that dated to 1689—disappeared from the office's files.

Around the same time, more and more settlers were finding the lake and seeking to derive a profit from it. By 1864, a trader and mail carrier named Solomon Barth was packing Zuni Salt Lake salt to Mormon settlements along the Colorado River and miners and ranchers across eastern Arizona. Settlement in the area accelerated in 1881, when Congress made a massive land grant to the Atlantic and Pacific Railroad in order to build a southern transcontinental railway, which would pass through Zuni territory. The grant gave the railway the jurisdiction to sell the lands to white settlers for grazing, logging, and speculation. Meanwhile, the government declared the salt lake a federal mine. Zuni lands were eroded and denuded, mined for coal and salt, and looted for saleable archeological objects.

Then, in 1898, Congress deeded all saline lands in New Mexico to the territory as part of its school lands allotment. This meant that the lands were required to be leased, and the royalties used to support a public university. The territorial government leased the lake to a settler named W. A. Rogers. The following year, sixty members of the Zuni, Acoma, and Laguna pueblos showed up at Rogers's encampment and told him that the lake belonged to them. They threatened to arm themselves if he kept them from visiting the lake. Rogers informed the local Bureau of Indian Affairs agent, who wrote a stern letter to the three pueblos stating that the lake now belonged to the territory, which had the right to lease it to whomever they wanted.

By that time, the tribes' offerings to the salt lake had largely been stolen by white visitors. A small settlement of Mexican-descended residents lived along the shores of the lake and mined and sold the salt—including to the tribes, whom they prevented from harvesting the salt using their traditional methods. The miners embarked on the lake in shallow canoes and scraped the salt from its bed using tools that Anglo observers reported looked like "forks of many tines." Settlers had made trails into the cinder cones, built dressing rooms, and begun to advertise the lake as a cure for rheumatism. Visitors swam in the lakes and desecrated graves.

The tribe complained, and in 1910, the BIA agent assigned to the Zuni wrote to the commissioner of the agency in Washington: "The Mexicans are crowding in on the south, the Mormons on the east and the Navajos on the north and west, and the Zunis are fearful lest everything be taken up and they crowded into their little reservation which is entirely inadequate to their needs." Eventually, the federal government sent a special investigator to the reservation, who concluded that its lands were indeed insufficient. The government added lands to the reservation in 1917 and again in the 1930s, the 1950s, and 1960s, for a total of 407,000 acres—still less than one-tenth of the Zunis' original homeland.

The new lands didn't include the salt lake, which continued to be invaded by settlers. During the late spring "salt season," landowners sent their hired ranch hands to the lake for several weeks of grueling salt-digging. In the late 1930s, the lessee I. N. Curtis used a bulldozer to build dikes around the lake and blocked the areas where freshwater entered it. He built an evaporation pond, a pier, and a pump house, and designed a floating mechanical harvester that resembled a rototiller—though it turned out that the device took on water and could not float when the lake was shallow enough for salt production. In 1963, a subsequent lessee from Los Angeles constructed new evaporating pans and attempted to harvest salt with a road grader, with plans to market it to highway departments, commercial water-softening businesses, ice cream factories, and lettuce farmers, who would use it to accelerate the plants' maturation. NASA and the USGS were also planning to use the maar to test moon-landing equipment.

During this time, Gregory Crampton, a historian at the University of Utah and a photographer and writer, traveled to New Mexico to document Zuni life. His photographs from the visits depict adobe and stone homes with roughly framed wooden doors, windows, and roofs, some with metal television receivers on top. Corn Mountain is visible in the background, along with dramatic arrangements of clouds whose flat bottoms recall Georgia O'Keeffe's famous "Sky Above Clouds" series. Goats graze the streets; families bake bread in earthen ovens and store the loaves in plastic laundry bins. The men wear bolo ties and cowboy hats; the women wear dresses with Mary Janes and socks, perhaps dressed up for Crampton's visit. At the time, most people spoke only Zuni.

In 1972, a man named Milford Hall from Eagar, Arizona, came to the office of Zuni governor Robert Lewis to inquire about delivering salt to a tribal member. When Lewis realized that Hall was the new owner of the lake, he asked for first right of refusal if he ever decided to sell. Hall was the owner of the lease, not the lake, but he didn't correct Lewis's misunderstanding.

A few months later, Hall wrote to Governor Lewis, offering to sell for $225,000, the amount his firm had paid two years earlier.

Hall stated that he had been producing four to six thousand tons of salt, "a minimum amount that could be produced annually," adding that an engineering firm had assessed the lake and promised output of up to 20,000 tons annually with certain investments, such as new evaporating ponds and new dikes.

Governor Lewis consulted with the Zuni Program Officer of the Bureau of Indian Affairs, who agreed that while the price was high, the lake's religious and historic significance meant that it merited the purchase if the tribe could find the money. By chance, they had it. They were in negotiations with the Tucson Gas & Electric Company for a right-of-way across the reservation, and decided to use the easement payout—the most money the tribe had ever had in its coffers—to obtain the lake. Only when BIA employees appraised the property did the tribe understand that they were purchasing a lease, not a fee-simple deed. Still, they moved ahead with the transaction, paying Hall a down payment in October 1972.

They immediately ceased all mining operations. The lake was not in good condition. It was filled with silt and tumbleweeds, and nearly dry due to drought. Oxidized, unusable mining equipment—misrepresented by Hall as in working order—was scattered around the crater. Yet when operations stopped, they learned that the terms of the lease required them to produce salt. They began paying the state the minimum required royalty, but knew it was only a temporary solution. "We must have the legal title to the land in order to secure this sacred shrine for our people and future generations of Zunis," Lewis stated.

New Mexico's land commissioner was supportive of the tribe's efforts, but his hands were tied by the terms under which the federal government had deeded the lake to the state. The tribe's only recourse was a land trade between the state and the federal government. Only Congress could approve such a swap. And there was one big problem. Technically, the tribe didn't have the right to request a land transfer from Congress.

During the early twentieth century, tribes brought numerous land claims before the federal government, and Congress handled them on a case-by-case basis. Eventually, legislators got tired of the claims and, in 1946, the federal government opened the Indian Claims Commission. The commission gave tribes five years to make any final claims; once the commission's hearings finished, tribes would not be allowed to make any more requests.

To win their cases before the claims commission, tribes had to demonstrate that land was theirs under a narrow idea of "Indian title," which required proving "exclusive use and occupancy," including "physical control or dominion" and "actual use" that was "continuous . . . for a long period of time prior to loss."

The claims were challenging to prove. When the Northern Paiute tribe alleged that the government had dispossessed them of an area spanning 50 million acres, stretching 600 miles north-south from the Owens Valley to southern Oregon and 300 miles east-west to encompass most of Nevada, the case came down to a debate over whether the Northern Paiute had been a unified nation at the time of its dispossession. As an expert witness, the anthropologist Julian Steward testified that the region's twenty-six bands had been connected by culture and language, by visiting one another "for dances and rabbit drives," and "by lack of local band ownership of hunting and seed areas."

The commission disagreed. They stated that "the aborigines' concept of 'ownership' is immaterial to our determination of Indian title" and that because the Northern Paiutes had been "completely lacking in any overall political organization," they were not, according to the commission, "an aboriginal land using entity." The decision reads like a slap in the face: not only did the US government dispossess the Paiutes of their lands, now they claimed that their nation—created by the government as the entity through which the tribe was legally able to make its claim—didn't exist.

Those tribes that were successful in their claims didn't receive land, but money. Ute, Southern Paiute, Chemehuevi, Mescalero Apache, Cheyenne–Arapaho, Comanche, Navajo, Goshute Shoshone, and the Pueblos of Acoma, Laguna, Zia, Jemez, Taos, and Santa Ana, and others received awards ranging from $300,000 to $35 million. In one famous case, the commission awarded the Sioux tribe of South Dakota $17.5 million for the Black Hills. The tribe refused the award; they only wanted their land. Now worth over a billion dollars, it remains unclaimed in the US Treasury.

The commission had closed two decades earlier, and the Zuni were one of very few tribes that hadn't made a claim. If some kind of wrongdoing had prevented them from filing one, their lawyer advised them, perhaps they could persuade Congress to make an exception. But first they had to figure out what had happened.

They knew that the tribal governor at the time, Leopold Eriacho, had an adversarial relationship with the BIA. He had complained that the Zunis were receiving bad representation from the agency, and wrote a letter asking that their agent be removed and "replaced by some western man who . . . knows something about livestock, farming and someone who knows a little of our way of life," adding, "we have had one Eastern Hotshot after another and who has done nothing but mess things up and keep us all upset." The BIA had responded by removing Eriacho from office and installing a new governor, Conrad Lesarlley.

At first, they assumed that Eriacho hadn't filed a claim because he wanted nothing to do with the agency. But Eriacho had also committed significant effort to recovering documents that established the historic scale of the Zuni homeland, including traveling to the National Archives in Washington, DC, to view old War Department records.

Tribal leaders did recall two *melika*—the Zuni word for non-Mormon Anglos—coming to the tribe to discuss land claims while Eriacho was governor. At the time, the tribal government did not have an office or a secretary; the council met at the governor's house. "Those of us who made our livelihood with stock worked at that occupation during the day and met at night to perform our council duties," explained councilman Sidney Neumeyah. "For that reason, there was hardly ever a time when we were all present at the Governor's house at any one time." So when "two white-men came to Zuni for the purpose of inquiring about the land claims," the governor asked them to wait for another date when all council members could be present. But, said Neumeyah, "we waited for the two Melika but they never returned."

Then, a clue surfaced. On April 11, 1951—at the very end of the claims commission's window and the day that Governor Lesarlley was installed in office—a BIA agent named Dewey Dismuke had sent a typewritten letter, signed by Lesarlley and the tribe's deputy governor and head councilman, to the BIA's regional director. The letter said that the tribe had no knowl-edge of the federal government having taken land from them and stated, "I wish to inform you that we do not believe that we have a claim to be filed."

Historian E. Richard Hart, who was conducting research for the case, went to Dismuke's home in Albuquerque to find out what had happened. The retired agent told him that he had felt that as a federal employee, it would have been inappropriate to suggest to the tribe that they file a claim against the government—despite the fact that giving such advice was part of his job. He also claimed that the tribe had had conversations about filing a claim, but that they had all taken place in the Zuni language, meaning that he couldn't understand them, although at the same time he claimed to know that the leaders had decided against submitting a claim.

It became clear to the Zunis that the BIA had failed them. They made their first appearance before the Indian Affairs Subcommittee of the Interior and Insular Affairs Committee of the US Senate on June 6, 1975, testifying in support of a bill that had two components: first, granting the tribe a waiver of the claims commission statute of limitations, and second, directing the Secretary of the Interior to execute a land swap in order to acquire Zuni Salt Lake and its surroundings. "We have come to Congress for justice for our people, because it is the last resort for relief," governor Edison Laselute told the committee.

Former governor Lesarlley stated, "I feel very badly because I didn't know what I was doing. These government people should have explained to me more fully and had my Tribal Council here and some of the Caciques might have been here so that they could explain it for the whole Tribe. I just hope and trust in God that the Congress of the United States will give us just consideration." Sol Ondelacy, the tribe's official interpreter under Lesarlley, added:

The letter which Governor Lesarlley signed was a misunderstanding, that's for damn sure. I can't blame anybody else but the BIA, who was responsible for the Zuni people and were supposed to give them the right facts instead of telling them it's not going to be any good if you try to put in a claim. I don't know how they approached Mr. Lesarlley or how he understood it, because I wasn't there to interpret it . . . I don't know who wrote the letter that Governor Lesarlley signed, but I can tell you for sure that he didn't write it himself, because he is about like I am. I can speak English, although broken up, but to write a letter in a White Man's words, in writing, I can't. I can write just common, ordinary letters to my children, but all broken up. Conrad is about the same way.

The committee recommended that the Senate adopt the bill, and they introduced it in February of the following year.

The Department of Justice and the Department of the Interior recommended against the bill. The Acting Assistant Attorney General went so far as to write, "the Government of the United States has in the past dealt very generously with this tribe. An Executive order reservation for the Zuni containing approximately 405,000 acres was established by Action of President Rutherford B. Hayes on March 16, 1877," ignoring the evidence showing that the reservation was a small fraction of the tribe's ancestral territory and was too small for their needs, as well as the fact that the 1877 allocation was in fact even smaller; most of the acreage was added later. Meanwhile, the BIA—part of the Department of the Interior—stated that despite documentation of the lake's historic and religious significance, they did not see why the federal government should have the responsibility for acquiring it for the tribe.

The bill would have to be reintroduced twice more. On May 15, 1978, Congress finally passed Public Law 95–280, which authorized the Secretary of the Interior to acquire the lake and 5,000 surrounding acres and hold them in trust for the tribe. It was almost a year to the day before the Mono Lake Committee filed their lawsuit: the two fights unfolded back to back, each unaware of the other.

As in the Mono Lake case, the decision took years to take effect. The Bureau of Land Management and the state of New Mexico had to conduct appraisals, find a suitable tract of land to trade, and wait for the other's approval at each step of the process. The tribe, still on the hook to pay the state for the mineral lease during the delay, grew impatient. Finally, on November 11, 1985, the Zuni regained title to the lake.

To me, the case marks an early victory of Land Back, a political movement considered to have been started in North America in 2018 by Aaron Tailfeathers of the Kainai Tribe of the Blackfeet Confederacy of Canada and which gained steam during subsequent pipeline blockades. I suspect that it didn't start sooner because of the Indian Claims Commission, designed as it was to foreclose all tribal claims. Indeed, many of the lands returned to tribes since 2018 have come from churches, utility companies, and land purchases—not from the federal government. But the Zuni defied that arbitrary rule. In doing so, they opened the door for all of us to think beyond the colonial geography in which we live.

It took me a long time to write about ranching because I could not make sense of the intense, contradictory emotions and politics it inspired in me. Sometimes, working in ranching didn't feel like part of the sequence of my life, but its richest and most concrete period, the emotional anchor that all my reading and writing sought, in some way, to achieve and to understand. My thoughts felt alien to the dominant environmentalist politics of the West, which frames ranching as a destructive force but never sees the rest of the colonial map.

While ranching provides the easiest image of the settler domination of the West, the problem doesn't belong to ranching alone. It's a problem of the entire settler West. It is even, perhaps, more insidious outside of ranching, where it is easier to avoid confronting the histories and power dynamics inscribed on the landscape. The politics that prefers that the West be a landscape instead of a country—the politics that has ascended as the West has become synonymous with outdoor recreation—ignores the fact that recreation is also a settler activity, and recreation lands are just as stolen as grazing lands. When I go camping and post photos of spectacular views to Instagram, I'm no less of a settler than I was moving cattle on horseback.

Toward the end of my time as a cowboy, I found a Western writer who helped me articulate these contradictory observations and feelings. William Kittredge grew up on his family's million-acre ranch in southeastern Oregon—a place that, he writes, "seemed in those horseback days to be endless." He lived there until he was thirty-three, when he left to study writing and eventually to become a professor at the University of Montana.

Kittredge knows innately all the beauty of ranchers' way of life. "Out in our land-locked, end-of-the-road, rancher valley, the air was bright and clean with purpose," he writes. And he mourns the disappearance of the world in which he grew up. "Imagine those shining snowy mountains against the sheltering endless bowl of clean sky," he says. "We will not see such things again, not any of us, ever. It's gone. We know it is."

At the same time, he knows those feelings to be the product of colonial domination. Even the idea of aloneness, the search for solitude in nature, is a product of settlement and empire. But the desire and nostalgia for it is also real. The brilliance of his writing comes from his ability to see the larger picture without casting scorn on those feelings. He critiques them at their source without invalidating them.

While reading the book *Owning it All*, a collection of essays about Kittredge's ranching past and other topics, I went on a walk with a student who is a member of the Haida nation, which spans from the Pacific Northwest of the United States to southwestern Canada. When I explained that Kittredge had left ranching at thirty-three, she asked, "What made him realize?"

I felt myself jump to a defense, wanting to tell her that ranching wasn't all bad. But I calmly replied that I didn't know. I hadn't finished the book yet. That night, I read the title essay, which discusses the profound changes that World War II wrought even on Kittredge's remote corner of southeastern Oregon.

"Despite the mud and endless hours, the work remained play for a long time," he writes, "the making of a thing both functional and elegant." Then, everything changed.

> . . . it all went dead, over years, but swiftly.
>
> You can imagine our surprise and despair, our sense of having been profoundly cheated . . . For so many years, through endless efforts, we had proceeded in good faith, and it turned out we had wrecked all we had not left untouched. The beloved migratory rafts of waterbirds, the green-headed mallards and canvasbacks, the cinnamon teal and the great Canadian honkers, were mostly gone along with their swampland habitat. The hunting, in so many ways, was no longer what it had been.
>
> We had come to victory in the artistry of our playground warfare against all that was naturally alive in our native home. We had reinvented our valley according to the most persuasive ideal given us by our culture, and we ended with a landscape organized like a machine for growing crops and fattening cattle, a machine that creaked a little every year, a dreamland gone wrong.

Following the instructions they were given, they had gotten too good at ranching, too good at controlling the land, at making it industrial. What had been something that felt in sync with the natural environment had become something so destructive that the Kittredge finally had to walk away.

She was right: he *had* had a realization. Making our conversation feel fated, in one of the final essays of *Owning it All* Kittredge travels to Haida Gwaii to write about Haida artists. I loaned her the book and told her about the coincidence. When she returned it, she told me that the essay was largely about her own family.

In many ways, the ranching pendulum has swung back from its industrial postwar ideology. The people I worked alongside thought in ecological terms. Amy had even trained on the ranch that was the setting of *The Last Ranch*, another surprise that felt fated. Still, my memories of the times when ranching didn't feel right are those when it felt like we were working against nature. The landowner sending us out with a USDA representative to put neon-pink pellets all over a prairie dog town on the south edge of the ranch, despite Nick and Amy's apprehension. The day that, while working alone, I found a pronghorn whose leg was caught in a barbed-wire fence, its face pathetic and the bone sticking out of its broken leg, and how I didn't know to look away when the Fish and Game agent shot it because I hadn't known what he meant when he said he was going to "dispatch" the doe. And branding—the major celebration of the year—for me was too loud, too rough, too male to enjoy.

In moving away from ranching and learning to see it with distance, I began to understand it as something I had to disavow on the long horizon. I began to see it as, in Cree writer Billy-Ray Belcourt's words, an "incantatory performance of an impossible subject position." It became a structure of feeling I could no longer inhabit.

These days, I think the only ethical position regarding land in the West is a desire to find a way to shrink the settler footprint, restore land to tribes, and re-imagine politics in whatever way such changes would require. We still need to recognize the authentic sense of place felt by settler residents, but we also must recognize that those attachments must be subordinated to other, more necessary, politics, including Land Back. We need to "worship an impossible future"—in Belcourt's words again—and gradually, collectively, make it less impossible.

Yet I couldn't have come into the convictions I hold now without the "becoming with" of ranching, without learning how it feels to be embedded in country and its more-than-human beings. Ranching is the reason I can understand what Métis anthropologist Zoe Todd means when she says that her parents' work in teaching her "about the land, waters, fish, berries, invertebrates and other beings of where I grew up was an instructive form of philosophy and praxis which imbued within me a sense of my reciprocal responsibilities to place, more-than-human beings and time." As much as ranching is an impossible subject position, in the way it recognizes deep engagement with country as requisite for living there, in the sustained commitment to place so rarely demanded of the rest of us, it sometimes feels like less of an impossible way of being than consumptive, indoor urban and suburban settler-colonial life.

Since leaving ranching, I've never recovered the sense of feeling part of the ecosystem that I experienced while working outside, all day, no matter what, measuring rainfall, estimating forage, doctoring calves, butchering. Looking back at my photos, the sensorial richness of the life overwhelms me. The clothes we wore, the tools we used, the smells of the animals. My leather boots braced against the stirrups, my straight-leg men's Wranglers stained with mud and horse sweat from gathering the horses bareback each morning, one of my favorite chores. My light-blue-with-white-polka-dots button-down work shirt. The thick leather reins resting gently in my tanned, calloused left hand. The silk paisley scarf around my neck, a birthday present from Amy. My perfect posture in the saddle. My strong, sinewy body. How happy I look. The embodied aesthetics of working country.

The whole time I worked as a cowboy, during my long stretches alone, I recited the Gerard Manley Hopkins poem "Pied Beauty" to myself.

Glory be to God for dappled things—
    For skies of couple-color as a brinded cow;
      For rose-moles all in stipple upon trout that swim;
Fresh-firecoal chestnut-falls; finches' wings;
    Landscape plotted and pieced—fold, fallow, and plough;
      And áll trádes, their gear and tackle and trim.

The easy gearshift of the Toyota, the cold quesadillas we packed for lunch, wrapped in tin foil and placed on the truck dashboard as, we joked, a solar oven. The resistance of closing a tight electric gate, the feeling of twisting wire around a century-old cedar post and of the rubber grip of my red-handled fencing pliers. Folding the fraying pasture map into a rectangle, sliding it into the sheer inner pocket of my Carhartt parka. My wintertime morning chore of driving to the cows' pasture and breaking the disc of ice floating atop their stock pond with an axe. The beauty of the country never getting old.

I still have it memorized, for when I need to transport myself back.

CHAPTER 6

A RIVER
PASSES BY HERE

Lake Texcoco, Mexico City, Mexico

I N THE COURSE LISTINGS, ONE CLASS JUMPED OUT AT ME: Queer Literature and Theory. I wavered before signing up. I slept on it one night, then another. It's an academic interest, I told myself. Anyway, I'm a feminist. Also, I thought, I'm in Mexico: no one is watching.

Officially, I'd come to Mexico for an exchange semester at the Universidad Nacional Autónoma de México while I studied for my PhD comprehensive exams. I needed fluency in Spanish for my dissertation research, and I wanted to study at the country's famous national university. Really, though, I'd come fleeing my breakup with Dylan. After living apart for two years, we had reconvened our lives in Berkeley, pursuing our respective doctorates: his in classics, mine in geography. But with each passing semester, the ways we wanted to grow apart were clearer: we argued when I wanted to go hiking or camping on weekends and he wanted to stay home; I felt that he didn't take my studies of the Southwest as seriously as his of Ancient Greece; and it was explicit, in our conversations about the future, that he did not plan to follow me to a future job.

Bored of talking about Plato and Pindar, and staring down a future as a trailing spouse, I had begun to feel that I was living in

the wrong life. I spent more and more time away from our apartment, drifting toward friends that were more mine than ours, most
of them queer of some stripe, people who I felt took my ideas and
intellect more seriously and with whom I could talk about ecology
and novels. But when I finally made the decision, it ruptured my
daily life and the future I had planned for myself. I quickly discovered that I wasn't going to make it through a rainy Bay Area winter
alone in the studio apartment that used to be crowded with Dylan
and his books, clothes, guitars, and mislaid pens. I'd dreamed up
my semester as a visiting student as a geographic cure, a way to
replace loneliness with freedom.

I got to room 322 early. Light pelted the classroom's south-
facing wall of windows. Mexico City sits at 7,500 feet in elevation,
under a fierce sun in a large endorheic basin so commonly referred
to as a valley that the federal hydrology service's official name for
it is the Basin of the Valley of Mexico. Even at 9 a.m., the room
heated up fast. The students took the chairs along the interior
wall, then those in the middle, then those baking in the light coming in through the window. Then, there were no more chairs. But
students kept coming. Some sat on the floor between the desks,
others behind the professor's table, others waited in the hallway.

César, the professor, showed up at 9:15 wearing pink hi-tops,
oversized rings, and a flowered shirt unbuttoned to reveal a light
blue jewel on a silver chain. He was my age; he'd only received his
PhD a few months earlier. I was charmed.

He called the room to order.

"Your first assignment is to bring fans," he said. He unzipped
his backpack, pulled out his own, and began to dramatically
fan himself.

Then he talked theory.

"Everything we're going to read has to be situated," he said.
"This theory is imported. It's from Europe and North America.

It's not *nuestra*, so it doesn't all apply. Mexico is different. Here, for example, there are very few roles in society for gay men. What can you be? An artist? A hairdresser? Maybe you could be gay and be a professor—but this gay?" He fluttered his hands to show off his rings, then put them on his waist and did a hip circle. "I assure you all: I'm the gayest professor in this faculty, in this university, and therefore in this country. So part of being here, in our beautiful classroom, is to say, yes"—another hip circle—"we exist."

Like Salt Lake City, Mexico City was constructed as a sacred metropolis on a salt lake. The Nahuatl-speaking Mexica people came to the Mexico basin in the early fourteenth century, led from the north by a hummingbird. When the bird took the form of an eagle and landed on a cactus growing out of a rock, they knew it was the sign for them to settle. The cactus the eagle had chosen was on an island in the center of Lake Texcoco, the largest of the five connected lakes—two fresh, three saline, totaling 424 square miles—that drained water from the forty-five rivers running down the valley's porous volcanic mountains. In Nahuatl, *México*, the name of the basin that lends itself as a metonym to the name of the country, means "the navel of the moon."

The Mexica built their city, Tenochtitlán, on the island. It was a lacustrine city: a city defined by a lake. They built causeways to the firmer land around the waters. Like the Mormon grid extending outward from its central temple, the main axes of Tenochtitlán radiated north, east, south, and west from the vertices of the pyramid at its center. The boulevards' sightlines provided an uninterrupted view to the ring of the valley around them. "The city stretches out: / radiating brilliance like the quetzal feather," as a Nahuatl poem describes it.

They understood the world around them through *teotl*—the

concept that sacredness manifests itself in the landscape, through occurrences like waterways, storms, and mountains. The physical world was, in the words of art historian Richard Townsend, "magically charged, inherently alive . . . with this vital force." Rainstorms were brought by Tlaloc, a temperamental god who needed to be placated with human sacrifices. Two riparian trees—the willow and the *ahuehuetl*, or Montezuma cypress, which has an enormous, ridged trunk and can live a thousand years—supported the heavens and had given birth to the world. Streams flowed out from the cave at the base of the trees' trunks, connecting the human world to the underworld just as streams connected the mountains to Lake Texcoco. Coatlicue, the goddess associated with groundwater, was wild and restless. "She killed men in water, she plunged them in water as it foamed, swelled and formed whirlpools about them; she made the water swirl . . . and sometimes she sank men in the water; she drowned them," writes art historian Barbara Mundy in *The Death of Aztec Tenochtitlan, the Life of Mexico City.* At Pantitlán, a site in the middle of Lake Texcoco where the water was particularly treacherous, sometimes even forming a whirlpool, the Mexica sent high priests to offer gifts to tame the water gods—including children adorned with green jade jewelry. Mundy writes: "Throats slit, their small and precious bodies were gently committed into Pantitlan's watery depths."

The Mexica built their city to mirror this sacredness of the world around them. The city was organized by *altepetls*—literally, "water hills"—which were springs associated with certain deities. These sacred hills determined the city's districts for governance, meaning that the idea of the altepetl took on the meaning of "the ideal socio-political community," as Mundy explains it. The city's four causeways divided it into its four altepetls.

The management of water was one of Tenochtitlán's rulers' most important tasks. They needed to control the summertime floods, which threatened to inundate the city; at the same time, they needed to bring fresh water for drinking and agriculture. Their waterworks expanded as their empire did. At first, they diverted the river that flowed into the city from the north during the rainy season, to prevent flooding. Then, they built a dam to solidify the border of the freshwater Xochimilco and Chalco lakes, so that when Texcoco overflowed, its saline waters wouldn't push upstream and contaminate their valuable drinking water. They constructed an aqueduct from the lush springs on the Chapultepec hill west of the city that became the city's royal entrance.

In 1449, Moctezuma I asked Nezahualcoyotl, the poet-engineer king of Texcoco, the kingdom on the other side of the lake, for help building a dike that would create a freshwater reservoir out of a portion of the saline lake. The result, Nezahualcoyotl's Dike, is believed to have spanned ten miles, separating the natural Lake Texcoco from the man-made Laguna of Mexico. "The process would have drawn on thousands of men and women to haul tree trunks, break stones, dig postholes, and make tortillas for workers," writes Mundy. They added a second, parallel dike in 1500.

The construction was accompanied by rituals in which priests performed the role of the deities. "As [the water] began to run toward the city, a man disguised as the goddess of the waters and springs appeared, dressed to impersonate the deity, in a blue garment," wrote Diego Durán of the opening of one of the aqueducts that brought water to the laguna. "His forehead was covered blue, in his ears were two green stones, another on his lower lip, and on his wrists he wore strings of blue and green beads . . . His legs were painted blue and he was wearing blue sandals."

I WAS RENTING A ROOM in an apartment in a gentrifying neighborhood north of the city center, with roommates who had grown up in the poor periphery of the city and studied the humanities for free at the UNAM. Luz, the master tenant, had grown up in Pantitlán, no longer the site of whirlpool and sacrifice but a neighborhood of tightly-packed cinderblock row houses near the airport. An architect my age, she was warm and outgoing; she had a pixie cut and spent her free time at dance classes. As soon as I met her, I wondered whether, if I put my mind to it, she might kiss me.

The Friday after I moved in, she asked if I wanted to join her to hang out at a friend's apartment a few blocks away. She put on a black cotton jumpsuit; I put on a men's extra-small button-down I'd gotten at a Goodwill in Palm Desert. The party was all gay men except for Luz, her friend Susana, and me. Alfonso, the host, asked me whether I knew why old American men were attracted to poor, young Mexicans. I told him I didn't. I sat on the floor talking with a philosophy student at the UAM, a university founded by professors too radical for the UNAM to serve the peripheries of the city. I told him I was taking queer theory and he laughed as I listed off the Mexican slang terms for gay men and the differences between them, which we'd discussed in César's class: *loca, musculoca, chacal, joto, gay.*

Alfonso and Luz came back from a run for beer, and the philosophy student and I joined the group. Susana was asking for a show of hands of who had been groped on the metro. Everyone raised a hand except me.

"Are you sure?" she asked me.

"Yes," I said. After all, I'd only been in the city a month.

"But there's also sexual violence among lesbians, right?" the philosophy student asked me.

"Yeah, it happens," I replied. I'd read about it.

"She says sexual violence also happens among lesbians!" he repeated to the group, but they weren't listening.

                                        SALT LAKES

When I got home, I realized it had been my first night as a lesbian.

The next weekend was a long one because of Benito Juárez's birthday. On Sunday, I went with Luz and her sister Criss—a "real" lesbian, which made me excited and nervous to be around her—to the city's antiques market. When we got back, I went to my room and tried to memorize dates from *The Spanish Frontier in North America*. I was tired and couldn't focus. I could hear people filling the living room. I didn't want to socialize, but eventually I got hungry.

As I made my quesadilla, everyone came into the kitchen to introduce themselves: Jess, Alma, and a tall, short-haired girl whose name I didn't catch. I joined them on the couches and someone put an Indio beer in front of me. Criss and the tall, short-haired girl were talking about living with their parents in "the depths of the city," as the girl called the periphery. I looked around. I realized they might all be lesbians, except Luz. There was Criss, of course. Then I saw Jess put her hand on Alma's leg. That left the mystery girl. I looked at her. How had I not noticed that she was gorgeous? Now I couldn't take my eyes off of her.

Criss was asking her about a recent breakup. I strained to hear, waiting for a pronoun. Her ex's name was "Alex," which was no help.

Finally, I caught an *"ella."* I decided to stare at her all night.

"I had to learn to take care of myself," she was saying, "not only to look out for her needs."

"That's such an important revelation," I said, feeling determined, "that you can care for yourself the way you take care of a partner."

"Has that happened to you with a man?" the mystery girl asked flatly.

I froze, wondering how I could keep up the lesbian pretense and also not lie.

"Come again?" I said, feeling ingenious; I'd left it ambiguous as to whether I hadn't understood or if I was politely saying, "What do you mean, with a *man?*" and giving her the opportunity to rephrase.

Instead, Jess interrupted.

"Who do you like? Men or women?" she asked.

I froze. "I'm a nun," I offered.

"For real?" everyone suddenly asked.

"No, no," I scrambled. "I just mean I don't have a social life."

Jess was resolute. "But do you like men or women?"

"Look," I finally said, trying to account for myself, stay honest, and keep my options open. I didn't even know whether I liked women, despite having committed to the bit. "I spent my twenties in a relationship with a man. But in the end, I couldn't do it anymore."

Criss patted me on the back affectionately. The mystery girl got up to get another beer.

"I'm only drinking today because it's a *puente*," she said. She must not have been listening, I thought. But then she brought me a beer, along with two for herself.

A friend of Luz's showed up with her boyfriend, who insisted that we move the party to his apartment, in the nearby Tlatelolco complex. I didn't really want to leave my apartment, but I knew I would go if the mystery girl came too.

We packed ourselves into two cabs and met at the 7–Eleven on the building's ground floor to buy chips and more beer. Everyone was already tipsy. We took the elevator to the twelfth floor. The doors opened onto an open-air foyer, and then we had to go up another flight of stairs. The apartment's living room was large, with huge windows without screens. Now the party was full of straight men playing guitar and congas, and it was boring.

I went to look out at the city. Luz joined me. She told me about how Tlatelolco was the site of the final conquest of the Aztec Empire. When the Spanish arrived in 1519, Tenochtitlán had a population of 150,000; Tlatelolco was a separate island, a breakaway altepetl founded in 1337. At first, Cortes had admired the lake-centered city. "To give your majesty a full account of all the strange and marvelous things to be found in this great city of Tenochtitlán would demand much time and many and skilled writers, and I shall be able to describe but a hundredth part of all the many things which are worthy of description," he wrote in October 1520.

His tone changed when he tried to capture it. The Mexica used the canals to their advantage, loading canoes with warriors and funneling the conquistadores onto the causeways, where they were easy to shoot. So the Spanish attacked the water infrastructure itself, including breaching Nezahualcoyotl's Dike so their boats, which they had assembled at Texcoco, could reach Tenochtitlán. The residents of Tenochtitlán fled to Tlatelolco. There, the Spaniards cut them off from water and food, leaving many to die of famine. They attacked the market, then set fire to the temple. "It burst into flames, the tongues of the fire rising high, crackling, and continually flaring up. There was weeping, there were tears among the Tlatelolcans, who expected that the people would then be plundered," one Indigenous leader narrated to Nahuatl-speaking Franciscan friar Bernardino de Sahagún.

The housing complex had been built in the 1960s in an urban renewal effort to clear slums and replace them with housing for the city's growing middle class. But it was most famous as the site of another massacre: desperate to quash dissent ahead of the 1968 Olympics, the government had opened fire on student protesters during a demonstration at the complex's main plaza, killing between 300 and 400 people. Then, after one of the buildings collapsed in the 1985 earthquake, killing everyone inside, everyone who had means left. The buildings fell into disrepair and became places to avoid.

"But it's changing," Luz said, gesturing to our hosts. "Young people are starting to move here."

Ordinarily, I would have loved the history lesson. But the eyes in the back of my head were fixed on the tall, short-haired girl. Luz caught on, and we went back to the couches. I resumed my staring. She had prominent cheekbones, and eyes that followed the angle of the ridges they formed. After a few minutes, Jess, who was sitting next to her, stood up.

"Sit here," she said, pointing to me.

The girl introduced herself as Patti. She complimented my green eyes and mustard hoodie and said she was a graphic designer who specialized in analog methods and loved to ride her bike around the city. "Can you believe someone so cool just showed up at my apartment?" I wrote in an email to a friend a day later. "I didn't have to do anything."

We agreed we were both cold and decided to switch hoodies. We put our hands on one another's thighs. The boys had turned on cumbia music, and she asked if I wanted to dance. I said yes eagerly, even though I didn't know how.

"You lead," she said. "I don't know how."

We flailed for a few minutes. Then she leaned in to kiss me.

I kissed her back for two seconds. Then I took her by the wrist into the kitchen, which was dark and empty.

"Sorry, I'm a puritan," I said.

I wrapped my arms around her, repeating over and over to myself that this was real. I was kissing a woman.

I felt her back and shoulders. Women's bones were so small!

I ran my tongue over her lips. They were so thin! And her teeth. Tiny! I felt her breasts against mine, and her small roll of belly fat. After a few minutes, we went back to the couches.

The party ended at 4 a.m. It was too late for Patti to get a cab to Iztapalapa, the borough where she lived with her mom. I told her she could stay with me. Luz and Criss hailed a cab, which drove us home without turning on its headlights.

We collapsed into my bed and started to spoon.

"It doesn't bother you if I haven't done anything with a woman in a long time?" I asked her. "A long time" was a lie, like saying you've already been a waitress on your résumé.

"It doesn't bother me," she said, which felt like generosity.

THE CONQUEST'S DESTRUCTION of Tenochtitlán's water infra-structure set in motion four centuries of efforts to drain the lakes. The Spaniards, mostly from arid Extremadura, had neither the skills nor the interest to maintain the lacustrine city. When they breached the Nezahuacoyotl Dike and its parallel Ahuitzotl Dike, the waters of the Laguna of Mexico began to flow back into Lake Texcoco, undoing the Mexicas' painstaking efforts to store fresh water. As the Spanish obscured the axes of the destroyed pyramid in favor of a Renaissance-style grid divided firmly into Spanish and Indigenous areas, the laguna dried into a smelly swamp, its canals too shallow for the canoes that had once been the city's main form of transportation. With the laguna dry, the Spanish used valuable drinking water from the Chapultepec aqueduct to irrigate their orchards, to clean prison cells, and to wash their ani-mals. They failed to maintain the canals and drains the Mexica had closely supervised, contaminating them with runoff from their slaughterhouses and tanneries and letting them fill with debris, sewage, and even corpses.

By the mid-sixteenth century, the city had run out of fresh water. At the same time, it was being flooded by heavy rains for the first time since the conquest. The three decades since the Spanish settlers arrived had been years of abnormally dry weather—part of the reason that they hadn't seen the point of maintaining the city's water infrastructure. Now, the Spanish viceroy sought advice from anyone who could help, and asked to see "ancient paintings" that showed how water had been managed in pre-conquest Tenochtitlán. The city's Indigenous council suggested rebuilding the Ahuitzotl Dike, while Spanish-descended engineer Francisco Gudiel suggested digging a tunnel to drain the water from the rivers and lakes on the north side of the basin, preventing them from reaching the city.

For the moment, the viceroy discarded the drainage idea. The dike was rebuilt, but the colonial government, upset that the Indigenous leaders had gotten their way, refused to give the workers the compensation they were owed, which was to be paid in corn. A page from the Osuna Codex, published ten years later, shows a drawing of a stone dike with blue water and waves on either side. In Spanish, the caption reads, "this is how they built the wall that protected this city from the laguna." In Nahuatl, it says, in Mundy's translation, "although the dike was built, they have not yet been paid."

Several years later, Antonio de Valeriano, an Indigenous leader who was the descendant of Nahua nobility and had received a European-style education in Spanish, approached the colonial government to request support for a proposal to ameliorate the shortage of fresh water in the city's Indigenous sector. He wanted to build another aqueduct from the springs in Chapultepec to a point in the city that is now the Salto de Agua metro station.

The new canal provided a lifeline to the city's Indigenous residents, writes Mundy, but perhaps more importantly, it also restored "the image of the altepetl once found in the Templo

Mayor, where a straight canal bearing freshwater from Chapultepec ran all the way into the temple." By returning water to the city,
he restored the city's sacred charge.

PATTI AND I texted each other and saw each other on the weekends. A few weeks later, she invited me to go to La Cañita. "A tropical queer spot downtown," she texted me, with a palm tree emoji.

I took the metro to Salto de Agua and turned the corner at the
city's civil registry building onto a dark, otherwise dead street in
the Doctores neighborhood. There was a thatched roof sticking
out from the entrance of the bar, and I could hear cumbia coming
from inside. It was a small space, decorated with odd relics. The
owners, a lesbian couple, were behind the bar serving *cevichelas*—
beers with a plate of ceviche surrounding them. Patti was sitting
on a stool next to Jess and Alma.

"*Hola, chula*," she said. I asked what that meant. "It's like
bonita," she explained. I felt happy. "I love this place," she said. "It's
like getting transported to the beach, and it's full of cool people."

When we finished our beers, we went outside so Patti and Jess
could smoke.

"This used to be considered one of the most dangerous neighborhoods in the city," Jess told me.

"What changed?" I asked.

"Nothing really," she said. "It still is. I'm sure they're paying *el
guante* to someone."

The bar filled up and we moved to the dance floor. One of the
owners was DJing. Everyone was moving. Everyone was beautiful, in a queer way. Everyone seemed so happy, packed into a tiny
space, in a bad neighborhood, in a violent country, in a misogynist
and homophobic world. I understood, suddenly, why so much
queer theory was about dancing.

"You're amazing, beautiful gringa," Patti said to me on the dance floor, sucking in a breath of air like she was about to dunk herself underwater. Instead, she kissed me.

I want to be gay forever, I thought, still not realizing that was in my power.

IN 1604, the city flooded again, with six feet of water. This time, the new viceroy didn't want to study ancient paintings or listen to Indigenous engineers; he wanted drainage. Spain's Royal Cosmographer suggested they try Gudiel's idea from decades before, of cutting a drain out of the closed basin. Sixty thousand Indigenous workers began digging the Huehuetoca Royal Channel, an open-air canal that diverted the Cuautitlán River away from the lakes. They finished in ten months.

When the city flooded again twenty years later and the canal failed, the government decided to rebuild and expand it, a process that took 150 years. They built a new tunnel to the state of Hidalgo, the Tajo de Nochistongo, which opened in 1789 and functions to this day. But it still wasn't enough to control the wild waters of the Mexico Basin. So the Spanish government called in someone who knew how to "measure and weigh the waters."

Adrian Boot, a Dutch engineer, was considered the world's leading expert in the science of drainage. He pointed out that continuing to drain the city would dry out the basin's clay soils—locally called *jaboncillo*, or "little soap"—and cause the city to sink under the weight of its buildings. So, he suggested a program of dikes, canals, and causeways—the type of waterworks the Indigenous rulers had promoted. The viceroy and Royal Cosmographer were unpleasantly surprised. They didn't want lakes or Indigenous engineering. As Alexander von Humboldt wrote in

1803, they "wished the beautiful valley of Tenochtitlan to resemble the Castilian soil, which is dry and destitute of vegetation."

WHEN HOLY WEEK CAME, my friend Nicolás offered to let Patti and me stay in his apartment while he was out of town. I proposed the idea to her, explaining the term "staycation." "Here we say '*Acapulco en la azotea*,'" she replied. Acapulco on your rooftop. She seemed excited, but I tried not to get my hopes up, as she sometimes canceled on me without much notice.

I told her to come pick up the keys with me on Sunday at 1 o'clock. I was worried she wasn't going to come. But at 1:45, she showed up on her bike.

"I didn't realize she looked like a model," Nico whispered to me in English while she was in the bathroom, before he headed off in an Uber.

"I can't stay," Patti said after he left. "I forgot that today is my brother's birthday."

I swallowed my disappointment and told myself she'd come stay another day. But when I texted her the next day, she said she was too busy with work. I slept on the futon without unfolding it and then went back to my apartment.

Soon, whenever we made plans, she ghosted me.

"Hey," I wrote her one of those days. "It hurts my feelings when you don't keep our plans."

"I fell asleep," she said.

"You had all day to cancel," I said. "It hurts my feelings that you don't communicate."

"I'm not responsible for your feelings," she replied.

Another day, I texted her to see if she wanted to get a beer. She said she was at the Punto Gozadera, a feminist bar, with her dog Hoppy, about to go home. I asked if she'd stick around for one more. She replied an hour later. "I don't have any battery left, I'm heading to Izta now," she said. I wanted to believe her, but by now I knew enough to be doubtful.

My phone rang at 2 a.m.

"Carito," she said, using Nico's nickname for me. "I'm downtown with Hoppy. Can we stay over with you?"

"I thought you went back to Izta," I said, half asleep.

"I call you in a crisis and you accuse me of lying," she said angrily. "No, Carito, it's better for me to stay away from you." I went back to sleep.

The following week, we made plans to watch a movie on Saturday night. She said she wanted to see me. But when I texted her to confirm, she said her mom was making her stay home with Hoppy; we'd have to see each other in the afternoon instead.

"But we had plans," I texted her.

"The truth is, I don't make plans with anybody," she wrote back. "The people who understand me accept that. I'm sorry that things don't flow with you. If you want to see me, I'll be at La Cañita." That was the opposite of not being able to make evening plans, I thought, but I didn't point out the inconsistency. I told her to let me know when she got there. I lay in bed watching TV on my computer, watching the clock tick from eight, to nine, to ten.

ONCE THE LAGUNA OF MEXICO dried up, politicians and engineers set their sights on Lake Texcoco. Into the mid-1800s, it still had water. "Around two hundred canoes traversed the basin's lakes and canals daily, supplying various foods, charcoal, firewood, and other goods from the surrounding hinterlands to the bustling urban markets," writes historian Matthew Vitz in *City on a Lake*. But like the laguna before it, the lake was gradually evaporating into a smelly swamp. The Spanish and their Creole descendants responsible for this desiccation considered the shallow lake insalubrious and rife with "miasma." Its main water source was, indeed, the city's wastewater. During the rainy season, the lake's "putrefied organic waste"—in the words of one visitor—flooded the city; in the dry season, it blew into the air as dust.

Though those in power had set their sights on draining it, others wanted to see Lake Texcoco restored. Humboldt had criticized the Spanish government for considering the lake "an enemy against which it was necessary to be defended," and argued, like others, that the floods should be controlled through dikes. The engineer Manuel Balbontín agreed, writing in 1870 that desiccation would "turn a place that should be gorgeous and productive" into a marsh "that would eternally sustain sickness."

Still, drainage won out. In 1886, after another flood, six-term president Porfirio Díaz created a commission to execute the most ambitious scheme yet. The Gran Canal de Desagüe, or Great Drainage Canal, an open-air channel designed by engineer Luis Espinosa, would drain the city's waters to Lake Texcoco. There, a new passage to the north would drain Texcoco's water to Lake Zumpango, where it would enter the six-mile Túnel de Texquiquiac and drain into the state of Hidalgo. He contracted British firms to construct the state-of-the-art system, which was inaugurated in 1900.

Complete desiccation hadn't been Espinosa's intention in designing the Gran Canal. Like Valeriano, Boot, Humboldt, Balbontín, and countless Indigenous residents of the city, he wanted to build an adequate sewer system and prevent flooding, while keeping the lake alive to provide green space and clean air to the city. But by the time the canal opened, the lake had been reduced to one-fifth of its surface area in 1519. Twenty years later, it only appeared after heavy storms.

"Three races worked on it, and almost three civilizations," writes Alfonso Reyes of the city's drainage in his 1917 essay

"Visión de Anahuac." "From Netzahualcóyotl to the second Luis de Velasco, and from him to Porfirio Díaz, the directive to dry the earth runs, it seems, unbroken. Our century found us still tossing the last shovelful and digging the final canal."

THIS WAS THE LAST STRAW, I thought. I texted a girl from Tinder to see if she wanted to get a beer. Criss had convinced me to sign up, to distract myself from Patti. "There are people that are soooo hot on there, *no mames*," she had said.

My date was a publicist for Grupo Modelo. She was cute and easy to communicate with. An hour in, like a switch in my brain, my need for attention from Patti suddenly turned off. Of course, she had her own problems: after our one-night stand, she texted me to apologize and say she'd only gone on the date because she was trying to flip her own switch. She needed to extricate herself from a complicated love triangle: her girlfriend had gone home to Argentina, and she'd started sleeping with the girlfriend's roommate, who had never dated a woman. But what she told me next made the disappointment worth it: "And you know, when you go out with a girl for the first time, it's all so different that you get attached really intensely, really fast. But you also don't have the experience to know who would make a good partner for you."

She wasn't talking about me, but I needed to hear it. I realized I'd suspended everything I wanted in a partner for Patti. She had represented the excitement of being queer, and I was still a fledgling. No theory had prepared me for that.

ONCE THE LAKE WAS DRY, the city's engineers had to figure out what to do with the thirty-acre Texcoco lakebed, whose dust clouded the city during the dry winter and spring months. In keeping with the Mexican revolution's redistributive goals, they decided to transform it into an agricultural area whose plots would be owned by small-scale farmers. They began re-engineering the lakebed: reshaping the surface so that all water drained to a central reservoir, "washing" the soil with groundwater to get rid of its salts, and establishing a tree nursery to cultivate salt-tolerant willow, tamarisk, and eucalyptus trees. They planted radishes, Brussels sprouts, lettuce, oats, alfalfa, and barley. When rinsing the salts from the soils turned out to be more complicated than they expected, they wrote to Utah's agricultural experiment station seeking help, and received a pamphlet in the mail in return.

Not everyone was happy. Local *campesinos* from towns like San Juan de Aragón, Ixtacalco, and Chimalhuacán who were used to harvesting salt, *ahuautle* (water-boatmen eggs), and other foods from the shoreline protested the disappearance of the lake and their limited access to it, now that it had become a federal experimental zone. Without water, the bird life that had once provided ample prey for hunting was also no longer abundant. Villagers from San Juan de Aragón wrote, in Vitz's translation: "this pueblo has sustained itself principally from the salt marsh and from the different products of Lake Texcoco . . . Now all those resources have been exhausted."

The government claimed that the agricultural program was in the campesinos' interest, but canceled it in 1946, letting the salts return to the soils and the lakebed become an expansive springboard for plumes that blocked out the sun, stopped traffic, caused airport shutdowns, and covered residents in layers of beige. Between 1923 and 1939, there were up to seventy-four dust storms in Mexico City each year.

"Geographer and essayist. I like to read and take bike rides," I typed into my profile after enough misadventures on

Tinder to realize that if I wanted a committed, calm domestic life, I needed to make it clear from the start. I'd been thinking about how many of my friends were older than I was, in their late thirties and forties. I appreciated their quiet, settled domestic lives, their social rhythms constituted by cooking dinner and taking walks and having conversations, their attention to the analog beauty of the physical world, and their senses of self-possession, surely hard-won over years but now internalized as their way of being in the world. It hadn't occurred to me before that I could live similarly, simply by finding a partner who already lived that way. I switched to Bumble, set the age bracket so that I would see matches up to age forty-two, and represented myself as honestly as I could.

Mariana's profile read, simply, "*Mariana, 40, arquitecta,*" a simplicity that seemed to me to exude confidence. She matched the image in my mind's eye of the person I desired: artistic with an aesthetic that veered masculine, cosmopolitan enough not to be deterred when I started talking about my doctorate or the books I was reading. In one photograph, a mirror selfie taken with an old digital camera, there were large framed posters of colorful constructivist designs in the background; in another, she sat in a modernist metal and wood chair—I later learned she had designed it herself—wearing a mechanic's jumpsuit.

She messaged first. "I hope you see this within twenty-four hours," she wrote in Spanish. I saw the message instantly, and laughed. It would never cross the mind of a millennial that I wouldn't see the message before the app made it expire. We moved our conversation to WhatsApp and made plans to meet at a tea shop the following weekend. She asked where I lived and told me I'd pass by her block on my way, so we agreed to meet on the corner and walk together.

When I arrived, she was wearing trousers with a large gray tartan, a linen shirt and a waxed canvas jacket. As I leaned in to kiss her on the cheek, a silver stud earring glinted through her loose curls. On the walk, she told me about Lázaro, her skinny, wild rescue dog, and how she would carry him up the three and a half stories of spiral staircase to her roof because he was too scared to climb them himself. She pantomimed it, spinning in three circles on the sidewalk with her arms in front of her, hugging the imaginary dog.

"I'll show you after our tea," she said.

At her apartment, a mixture of awe and envy came over me. She had renovated a garage and servants' quarters into a cozy loft-style space. The living room, the former garage, had a brown tile floor, the metal chair from her profile, and pine bookshelves resting on a blue metal structure. The chairs at the round kitchen table looked, from afar, like simple classroom chairs, but from closer up I noticed that their metal frame had an unusual boomerang style, and the seat and the backrest were cut from planks of a beautiful, reddish dark wood. She had designed it all. "I think it's nicer when everything in your house is handmade," she explained, as though that were an ordinary way of furnishing one's home.

I realized that all the homes that had sparked in me this pleasurable envy—the deep, clamoring wish to inhabit them—had been queer ones. When I was twelve, my brother and I had been child extras in the Opera Colorado production of *Tosca*. During the week of performances, a sign had appeared on the bulletin boards backstage: "CAST PARTY AT CHADWICK'S." We showed the flyer to our mom, asking if we could go. Thinking Chadwick's must be the name of a bar, she flagged down one of the singers to ask whether kids were allowed.

It turned out that Chadwick was just a member of the opera chorus. He lived in a brick duplex near Cheesman Park. Inside, the small living room was packed with middle-aged men dressed in their black performance attire. I suspected we were at the home of someone gay—it was the opera chorus party, after all—but whenever we asked our mom if someone was gay, she always denied it, even when it was true, so I couldn't ask. I went downstairs to use the bathroom, where I saw two framed posters of condom advertisements in a European language I didn't recognize with images of gay men having sex. I admired the small, simple bedroom, which had a mattress on a low frame and more art on the walls, and then went up to join the party, flush with a desire I couldn't articulate—the desire to live in one of those small brick duplexes just like that, listening to Puccini and drinking cocktails with my friends.

Then there was my graduate school friend Chris, who lived with his partner in a duplex in North Oakland. Volumes of Lorca and Bishop sat on cinderblock bookshelves and the new editions of *Environmental History* and the *Annals of the American Association of Geographers* lay on the coffee table. Treasures and trinkets given to them by friends, including the birthday card I'd drawn for Chris, decorated the rim of molding that encircled the simple, linoleum-floored kitchen. They fed the neighbor's cat and kept a vegetable garden.

Simply spending time in the house, I could tell they weren't following the same teleology of getting married, buying a house, and having children that I had felt Dylan and me accelerating toward. I felt envy, but also enigma. "I'm against marriage, I'm against monogamy, I'm against ownership," Chris had said with an out-of-the-blue declarativeness one morning when I told him that Dylan and I were going to a wedding that weekend. I spent weeks puzzling over what commitment must feel like in a partnership that wasn't focused on cementing itself through milestones.

Then, reading the collected poems of Frank Bidart, I found an answer: "*Being* is making: not only large things, a family, a book, a business: but the shape we give this afternoon, a conversation between friends, a meal." It was a description of what I loved about spending time with Chris—how a few hours with him felt like entering another world, one full of attention and sensory pleasures.

Over time, living with Mariana, I came to refer to this as the "queer quotidian," an atmosphere that derived its sense of meaning not from looking to the future and fulfilling others' expectations for success, but from its attention to the depth of each moment and each object. The queer quotidian meant delighting in fabrics and good food while also keeping a principled commitment to simplicity and the well-being of others. It meant finding pleasure in secrecy, transgression, otherness. Inside it, time was not a linear progression but a wide, constantly reconfiguring container of experience and pleasure and, sometimes, struggle. Mariana, I learned, had her own term for it: "*viven bonito*," she would say after we visited certain people.

In the morning, after a lot of mescal, I woke up in Mariana's bed. I opened my eyes and saw her button-down shirts hanging from the clothing rack. I smiled at her soft-butch style.

"A naked woman in my bed!" she said in feigned surprise. She pulled my arm across her body and kissed it. No one I'd been on dates with before had ever done that. And certainly no one had had handmade built-in shelving. She was living the queer life I coveted, and she wanted someone to share it with.

We made plans for a second date, to go to the UNAM's sculpture garden the following Saturday. We held hands on the bus to the garden. Afterward, we got coffee at a bookstore and she told me about her sculpture projects. For years, she had been sculpting with salt, mixing it with a tiny bit of concrete for structure. I told her I was working on a book about salt lakes. It was a strange thing to have in common, we agreed. She showed me images of salt flats I'd never heard of: in Mauritius, Bolivia, Baja California. Then she showed me an image of a set of salt cubes she had made. Each varied from the next, sections missing or small pieces of wood added. They looked striking all lined up, their perfect textures and shapes like a minimalist model of a neighborhood or the auto-constructed houses common in Mexico City's periphery. At her studio, she offered one to me to unwrap. "You can lick it," she said.

As Boot had predicted, the city was sinking. It had developed a nonsensical metabolism: in the obsessive effort to get rid of excess water, the government had simultaneously worsened the city's dust and flooding problems. Because they sent away all the city's rainwater, they had to rely on groundwater—to this day, over 60 percent of Mexico City's water is extracted from the subsurface. But without rain and runoff recharging the aquifer, the weight of the city compacted the soil beneath it, sinking into the pores that once held water. In contemporary Mexico City, people describe the *centro*'s colonial-era buildings the same way they do crooked teeth: *están chuecos.*

By the 1940s, the city center had sunk below the Texcoco lakebed—geologically, the lowest point in the basin. That meant that the Gran Canal, which relied on gravity to drain floodwater and wastewater out of the city, no longer functioned. So, though Alfonso Reyes thought that his generation had seen the last shovelful thrown and the last canal dug, the desagüe wasn't over.

In March 1967, Mexico City's public works department began building a new central drainage system—one that would beat the city's subsidence. Mariana's dad, a civil engineer, worked on this last-ditch effort, called the Deep Drainage System. They dug up the city, excavating 3.5 million cubic meters of soil, and installed tunnels a mile below its surface, deep enough to be in a different soil layer, below the easily compacted lake clay. Storm drains across the city took rainwater to shafts that fed the tunnels, which converged on the Emisor Central—the "Central Emitter," where the city spewed some 45,000 gallons of water per second through a massive tunnel parallel to the nineteenth-century Túnel de Texquiquiac. It was the largest public work the country had ever undertaken, employing 30,000 people. Mariana's dad told a story of being given a single day to dig up Paseo de la Reforma, the boulevard that runs through Mexico City's financial district, lay piping across it, and close it again. In our bathroom hangs a certificate presented to him commemorating the "effort" and "dedication" he contributed to the project's realization.

Ironically, the project made flooding worse in certain areas of the city. During the summer monsoons, the drains and shafts fill too quickly to handle all the water rushing into the tunnel below the city. To avoid overwhelming the system, engineers close the drains in certain areas of the city, allowing them to flood. On top of that, the improved drainage means that even less rainwater infil-trates the aquifer, depriving it of thirty inches of precipitation a year, more than falls on London.

W HEN THE PANDEMIC CAME, I hastily moved into Mariana's apartment. We spent long mornings in bed, drinking coffee and snuggling with Lázaro. We made the small space ample—I went to my desk in the living room, she went to her makeshift woodshop on the roof. We traded off cooking lunch.

More than anything, Lázaro's need for a walk each morning and evening structured our time. On an evening walk ten days in, Mar-iana led Lázaro and me to the intersection where Avenida Insur-gentes, the longest street in Mexico, crosses the Miguel Alemán Viaduct. She pointed to a concrete and metal tube that divided the two lanes. Inside of it, she explained, was a river—the Río de la Piedad, which had flowed enclosed since 1942. On top of the divider was the *ecoducto*: a 1.6-kilometer walking path lined with plants that help filter the water before it drains into the tube. "*Aquí pasa un río*," read a placard. A river passes by here.

In a previous apartment I'd lived just blocks away from the ecoducto, but had never visited it. I noticed the smell first: herbs cut through the smog, sewage, and burning trash that otherwise overwhelm Mexico City's air. The path was lined with lavender, rosemary, citronella. Mariana broke off a piece of lavender and stuck it through the buttonhole of her shirt.

By then, I'd started reading a collection of essays by Reyes that included "Visión de Anahuac." "In the middle of the salt lake sits the metropolis, like an immense flower of stone," he wrote. I'd known that Mexico City was built on a lake, but not that it was saline, nor that the city sprawled over an endorheic basin like those of Utah and Nevada and eastern California.

"Did you know Texcoco was a salt lake?" I asked Mariana when I read the essay, hardly believing that, after seeking out salt lakes from California to Kazakhstan, I'd been living—studying, going dancing, bicycling, taking the bus, kissing women, writing, writing about salt lakes—on the dry bed of one, for a year and a half, without knowing it.

"No," she, the *defeña*, the engineer's daughter, the salt-artist, said. "I had no idea."

AT THE SAME TIME as Mariana's dad was working on the Deep Drainage System, one of his professors from the UNAM was solving the problem of Texcoco's dust storms. Nabor Carrillo was son of Julián Carrillo, a renowned Mexican composer who created a theory of music called Sonido 13, believing that the future was in microtones, or notes with less space between them, creating scales of thirteen notes instead of twelve. Nabor, an engineer, studied soil science at Harvard, and eventually became the UNAM's rector.

After hearing of an idea from the early 1950s to re-flood the city to solve its problems of dust and subsidence, Carrillo became determined to take up the mantle. He came up with a plan: he would use treated urban wastewater to recreate a large portion of Lake Texcoco. He'd pump groundwater from underneath the lake to deliberately sink the elevation of its bed so that it would once again be lower than downtown Mexico City, and so that during the wet season, water would drain into it instead of flooding the city. He'd dig canals throughout the parts of the lakebed that weren't flooded, so that their water would drain to the reconstructed lake, too, and he'd reforest the remaining dry land.

He got approval and a budget. Though he died unexpectedly two years later, in 1967, the project continued. Ditches were dredged and embankments constructed; 9,000 hectares were planted with tamarisk and saltgrass, and four reservoirs of treated wastewater covered 3,000 hectares of the lakebed. The largest, named for Carrillo, is visible out the airplane window as you fly out of Mexico City, a rectangle with rounded corners like a shape you didn't mean to draw in Microsoft Word.

"He was a brilliant engineer," Mariana's dad told me when I asked him what he remembered about Carrillo. "But then he went and studied *en el gabacho*," using a disparaging nickname for the United States. "When he came back, he'd lost his sense of humor. He wasn't Mexican anymore."

TODAY, you can visit the Texcoco Lakebed. After a years-long fight over the possibility of turning it into an airport, marked by campesino activist fronts that demonstrated by swinging machetes in the air, as though chopping the airplanes out of the sky, it opened in fall 2024 as a park. To many visitors, the expanse of brush and saltgrass is probably boring, even disappointing. But for those who can read the landscape, its layers are evident. You drive along the Gran Canal and enter the federal zone demarcated by Porfirio Díaz. You see the rows of tamarisk planted and the canals dredged by Nabor Carrillo's teams. You pass through the tall chain-link fence once meant to enclose an airport and drive on roads meant to handle thousands of cars' worth of passenger traffic, whose parking lots have been turned into the activity areas of the new park. Lake Nabor Carrillo has a new embankment for birdwatching.

It's far from a natural landscape. Every square foot of it is engineered. Yet it nevertheless feels like a way of preserving the wild spirit of Lake Texcoco, the lake four generations of settlers tried to kill, which would never let them forget it was still alive.

You could call it a queer landscape. As a perspective on landscapes, queer ecology is often used to denote places that are degenerated, unruly, and inhabited by unusual, improbable groupings of species. Early accounts of such queer ecologies drew connections between the shared wounds of bodies affected by AIDS and those of degraded landscapes. Filmmaker Derek Jarman's book *Modern Nature*, for instance, is a diary of the garden he planted and tended in 1989 and 1990, at a seaside fisherman's cottage in the shadow of England's Dungeness B nuclear power station, while dying from AIDS. It's called "modern nature" because there's no pretense of it being pristine or virgin; it's a nature that belongs wholly to the modern world.

Around the same time in London, an abandoned and overgrown cemetery called Abney Park became a popular cruising site for gay men. The "unruly" nature and the unruly socializing benefited each other. A mycologist even observed that cruising improved the

                                                    SALT LAKES

mushroom species diversity of the site, by helping spread spores to new areas. In the 1990s, the once-forgotten site was recognized as a nature reserve. "The saprophytic gloom of the cemetery provides an ideal habitat" for fungi, wrote geographer Matthew Gandy in a 2012 essay, while "rare beetles thrive on rotting wood, a remnant fauna of moths and butterflies persists . . . and the woodland resounds to the sound of owls, woodpeckers, and other birds."

Lake Texcoco, too, is modern, unruly, and degenerated. It has been through everything. Yet in managing to survive, it has kept Mexico City a lake city, if in a changed form.

Queers are sufficiently experienced in the frustrations and disappointments of the world to understand that ideals of Edenic nature don't exist, but also practical and creative enough to make beauty in places others have left behind. The idea of modern nature teaches us that such landscapes—those that humans have intervened in and abandoned, those where wildness is now tangling the straight lines imposed by engineers—can also be rich ones. Queer ecologies celebrate the need to not manicure spaces so that they will radiate orderly, pastoral, and inoffensive visions—and teach us not to dispense with those places that seem to have irredeemably changed. Seeing our surroundings through such a perspective helps protect biodiversity and diverse human experiences alike. When we experience the joy and meaning of the natural world in places that feel eerie or imperfect, it shows us, in a bodily way, the need to protect marginal spaces and ways of life.

As Mariana and I learned to live with each other, soaking into the queer quotidian, we also began to find the lacustrine ecology of the Mexico Basin metropolis. We harvested lavender from a viaduct to make tea. We made pilgrimages to the old canals. We identified on street corners the *quelites*, or wild greens, we cooked at home. We planted tomatoes in buckets on the rooftop. We saw how our domestic patterns became parallel to those of a canine.

Many of our daily rituals revolved around water. Kneeling on the patio in front of the water heater and opening the gas valve to light the pilot light, and the *whoosh* when it caught. Waiting twenty minutes before showering. Adding three drops of iodine to the tap water, then boiling it. Always ending up drinking it warm. Mixing in hibiscus and sugar and still calling it water. The bikes welded with platforms to deliver water jugs. The harsh buzz of the pump bringing water from the underground cistern to the tank on the roof. Planning summer days around the afternoon downpours. The altitude that made the dishes dry so quickly. My skin so dry it felt too tight.

The city's lacustrine ecology wasn't gone at all. It took entering the queer quotidian to find it, itself a queer ecology, damaged and shunned, revealing itself only to those who appreciated it. Training our attention on those small daily tasks, making each one contemplated and meaningful, it allowed us to envision another type of city—one in which a sign saying *a river passes by here* would be unnecessary and redundant, because you could follow it all the way to its lake endpoint, the water visible and clean and alive. We wanted to live like that forever.

III.
SALINE FUTURES

CHAPTER 7

THE ECOSYSTEM THAT LAWSUITS MADE

Owens Lake, California, USA

THAT FALL, MARIANA AND I MOVED TO THE UNITED States. I'd been offered a semester-long position teaching writing at a college in eastern California, in a valley adjacent to the Owens and not far from the Mono Basin. We transported ourselves from one world defined by salt lakes—the Mexico Basin—to another, the Great Basin, the region I'd loved at first sight for the way it combined the color palette of the remote West with the comforting feeling of being enclosed on all sides by mountains.

On the drive from Denver, I wanted to stop and see Great Salt Lake. On previous trips, I'd always been in too much of a rush to add the extra time. Now, the air blurry with smoke from fires in California, Mariana egged me on as I turned off I-80 at its southwestern corner, crossing a set of railroad tracks and following a dump truck, ignoring the private property signs. At the end were mountains of salt belonging to Morton. I was as amazed as I had been at the Aral Sea. Finally, with the loading dock in sight, we turned around.

The following day, we went to the Bonneville Salt Flats, where we kicked a ball for Lázaro to chase on the white ground, the salts clumping like slush. It was a strangely sublime evening that felt like a snow day, even as the sky was lit orange.

Through the fall, it seemed that every weekend we camped next to a new playa or lake: Deep Springs Lake, Fish Lake, Saline Valley, Mono Lake, Badwater Basin, Searles Lake. So when the time came for us to get married, it seemed it only made sense to do so at the one we hadn't visited—Owens Lake.

The night before we did, I invited my colleagues over for drinks. The physics professor who lived in the other half of my faculty duplex asked what our plans were.

"You know that's not a real lake, right?" he asked when I told him.

Of course I knew. That was part of the point. Owens wasn't much of a lake any longer, but the dry bed of one, long desiccated by water diversions. After years of dust storms and mitigation efforts, it looked less like a body of water than a technical diagram, drawn and quartered by litigation and scientific intervention. But it felt fitting. Like the lake, our marriage bore a permanent and painful imprint of bureaucracy.

PLEISTOCENE-ERA OWENS LAKE formed eleven thousand years ago, when the huge glaciers that covered the Sierra Nevada melted and plummeted down the Owens River to the lowest point in the valley. It was over 200 feet deep and 200 square miles in area. That massive lake evaporated over time into Owens Lake, which covered 100 square miles and was slightly salty, mainly because of sodium bicarbonate, or "soda."

Then, settlers compressed geologic time into a matter of years. After the US Army murdered or removed many of the region's Native Californians during the Owens Valley Indian War, white homesteaders began diverting water from the Owens River for farming and ranching. Then came Los Angeles and its aqueduct. By 1924, the lake was completely dry. The salts left behind precipitated into a crust of trona, burkeite, halite, potash, potassium, chloride, and borax.

And dust. Since salt lakes collect not only rivers' water but also the silt they carry, their beds have a layer of fine particles that, once uncovered, is vulnerable to being picked up by wind. By the 1990s, the bed of Owens Lake was releasing up to 76,000 tons of dust each year. It had the highest measure of particulate matter in the United States, 123 times the EPA's limit, and included toxic matter such as arsenic and cadmium left over from mining. Residents of nearby towns kept gas masks and oxygen tanks in their homes. They had bloody noses and irritated eyes and lungs. Some died of lung cancer and emphysema. Many say they still have chronic respiratory problems.

The dust has a geography, a topography, and a season. At Owens Lake, the eastern crescent of the lake emits the most, because some water still pools and evaporates there. In the spring and fall, it dissolves the dry crusts of the previous season and then wicks the loose salts in the soil to the surface, where they bloom into delicate new crusts. The crusts are so short-lived that they only form tiny crystals, as Jay Owens writes in *Dust: The Modern World in a Trillion Particles,* that geologists refer to as "puffy" or "hairy." The winds blowing off the nearby mountains pick up the dry ions and light, clay soil, and fill the air with them. Since the center of the lake stays dry all year, meanwhile, it's held firm by larger crystals. And during the summer, even the delicate crystals at the edge of the lake harden into a crust that the wind can't kick up.

Great Basin Unified Air Quality Control District, the agency responsible for enforcing air quality rules in the Eastern Sierra region, started monitoring the air quality at Owens and Mono lakes in the late 1970s. In 1987, the Environmental Protection Agency set a standard for "fugitive dust"—tiny particles smaller than ten micrometers in diameter. The agency began trying to build a legal case that would apply the standard to hold the Los Angeles Department of Water and Power accountable for the dust. But the Ninth Circuit Court of Appeals declined to hear the case. The beds were, in the court's eyes, natural sites.

Then, Congress rewrote the Clean Air Act. In the revision, dust from so-called "natural sources" that had been altered by human activity could be held to emissions standards. It even name-checked the Eastern Sierra, writing: "The term anthropogenic source includes sources that are indirectly created by human activity as well as those that are the direct result of such activity. An example of such a source [is] the dust storms that are generated from dry lake beds at Owens and Mono Lakes in California."

With the law finally on its side, in the early 1990s the air pollution control district sued the Los Angeles Department of Water and Power over Clean Air Act standards. The big-city utility tried to starve out the rural regulation agency, filing seven countersuits. But in 1997, the courts decided in favor of Great Basin. They mandated that Los Angeles provide dust mitigation on an initial thirty square miles of lakebed.

By now, after years of litigation and engineering, the lakebed has better air quality than most cities. There are only around seven non-compliance days per year, and their particulate counts are tame compared to what they once were.

"We call it a success story, because we've successfully controlled the dust," Phill Kiddoo, then the Air Pollution Control Officer for Great Basin Unified Air Quality Control District, told me. "But you could see it as a failure, because we lost the lake."

Phill's job was to make sure Los Angeles actually did the dust remediation efforts the courts required. I met him a year and a half after Mariana and I got married, when we returned to Owens Lake to spend the day with him. We met at the Lone Pine Film Museum and joined him in a white Toyota 4Runner with govern-

                                    SALT LAKES

ment plates. It was March, and after a winter of blizzards there was snow on the mountains all the way down to the valley floor. The Mono Basin, just a few hours north, was snowed in. But where we were, it was warm enough that I wasn't wearing a coat.

Phill was an Eastern Sierra local, and he looked it. He wore a short-sleeve button-down shirt, forest green pants, and sportsman's sunglasses. Though nearly fifty, he looked younger. He was born in the resort town of Mammoth and lived there until he was five, when he moved to Bishop, the Eastern Sierra's largest town. After living in Santa Barbara for five years and Mexico for one, he married another Owens Valley local, and they moved back to Bishop to raise their kids. At first, he worked for federal science agencies, wanting to be outdoors. "I was born an artist, and then I become a scientist in college," he said. He never thought he'd take an office job, but the seasonal federal gigs weren't steady enough to support a family. Once he became the boss, he was on the other side of his old conundrum: it was hard to predict which of his staff would stay. "We call it the East-side résumé—having like seven employers in five years," he said, smiling at his invented term. "That's just how it is, because all the work is seasonal."

Our first stop was on the west side of Highway 395, which runs north–south down the Owens Valley, at the aqueduct. It was the diversion point for the "main line," the eighteen-inch pipeline that Los Angeles built to connect the aqueduct to the lakebed after the first court order in 1999.

After the 1997 ruling, LADWP had three years to comply with the court's mandate. The agency dragged its feet. When it finally accepted that it was on the hook, it had to work fast or it would incur a sizeable fine. It didn't want to give up its export water, but flooding the lakebed was the quickest way to control the dust. It quickly built the main line, which travels under the highway and onto the area it was required to remediate. There, it built rectangular ponds for shallow flooding.

"I think they hadn't realized yet that they would be here forever," Phill said of the last-minute work. No one realized what they had gotten themselves into, including the air quality agency. No project like it had ever been undertaken. "With most projects, you know what the impact is before you start, so you can plan the mitigation," he added. "We had to figure out how to mitigate by trying a lot of things that didn't work."

MARIANA AND I had decided to get married on one of our quiet quarantine afternoons in Mexico City. Waking up from a nap, she called me to the bedroom to lie down with her. She had been chatting over WhatsApp with a friend in Los Angeles. "Ernesto says if we want to move to the US next year, we should start applying now," she said.

I thought about it for just a few seconds. We wanted to stay together, to be able to live together. Marriage was the way to do that, since we held different passports.

"I think we'll be really happy," I said.

"We already are really happy," Mariana replied.

I downloaded the application for a K-1 visa, a single-entry visa designed to allow its recipients to enter the US and marry a citizen within ninety days, then apply for an "adjustment of status" to receive a green card. I spelled out the names of Mariana's parents and their birthplaces in my best all-caps handwriting and checked the addresses of her past jobs and apartments. Since all the passport-photo storefronts were closed due to the Covid lockdown, a friend who worked as an art photographer took photos of us against the white backdrop at his studio. Sitting at his computer afterward, I read the specifications from the application

                                                SALT LAKES

instructions, translating them into off-the-cuff Spanish: "Head height should measure 1 to 1⅜ inches from top of hair to bottom of chin, and eye height is between 1⅛ to 1⅜ inches from bottom of photo."

"Gringos are fucking maniacs," he said as he fiddled with the ruler tool in Photoshop.

We sent the paperwork via DHL to a facility in the exurbs of Dallas. Since it was a fiancée visa, we didn't yet have to prove that we loved each other—only that we had "been in each other's physical presence at any time during the two years immediately preceding the filing of this petition." It felt an absurd thing to prove, given that we shared an address and slept next to each other every night, but I nevertheless printed and copied every form of evidence the agency listed as acceptable: passport stamps, plane tickets, a photo with a handwritten caption identifying its date and place. A friend and my sister each wrote affidavits attesting that they had been in our shared presence.

Eight months later, in mid-February, we received a piece of thick paper with an engraved letterhead reading "The United States of America" in a font similar that on a dollar bill. The I-797 "Notice of Action" informed us that the USCIS office in Texas had approved our petition and had forwarded it to the National Visa Center in Portsmouth, New Hampshire. "Processing should be complete within two to four weeks," the notice said. After that, I inferred, we'd be contacted by the embassy.

When no notification had come by early June, I called a lawyer. He told me not to worry: he explained that embassies had closed for a year because of Covid and said we weren't going to get interviewed anytime soon—probably not for another year. The good news, he said, was that it wouldn't be a problem for Mariana to come with me to California in the fall on her tourist visa. We'd go for the semester, we decided, and then return to Mexico

City to wind down our lives while we waited for the immigration authorities.

We crossed the highway in the 4Runner and drove on the dirt road that follows the main line. As we entered, there was a sign that read MITIGATION AREA, HEAVY MACHINERY.

"I think they want to scare people off," Phill said. "Sometimes it seems like LA has an interest in people not coming here." He pointed to some metal trash cans alongside an information panel set up by the Bureau of Land Management. "I can't imagine they've ever had to empty those."

There were LADWP trucks driving on the roads between the dust-control cells, looking like toys in the vast beige-colored landscape backed by the Sierra Nevada's gargantuan eastern face. Around 10 percent of LADWP's workforce is in the Owens Valley, Phill explained; they handle day-to-day management of the aqueduct, run the town water systems for Big Pine, Lone Pine, and Independence, and have a vegetation management team responsible for pulling out tamarisk, pepperbush, and other invasive species.

On its way to the shallow flooding areas built to comply with the initial 1997 settlement, the road took us through the delta where the Owens River once dumped into the lake. A tall, sheer cornice, cut by water, overlooked the dirt road. The scale was difficult to fathom. The volume of water necessary to carve such a shape would have been more than enough to catch the SUV and carry us along with it.

Around the delta and the rest of the former shoreline, springs bubbled up from the ground and poured water into what was once the lake. "Nature doesn't know it's not a lake anymore," Phill said.

At the shallow flooding cells, meanwhile, water spouted out of "bubblers" and spread over the lakebed in a thin sheet. The cells had to be kept 75 percent wet, Phill explained; his staff checked them every five days using satellite imagery. When they were created, they were an unexpected success. They created a habitat where algae and brine flies flourished, providing food that attracted birds, which congregated around the bubblers to bathe and cool off. In 2002, the National Audubon Society declared the lake an Important Bird Area. Survey teams organized by the Eastern Sierra Audubon Society counted over seventy species in a single day, including fifteen thousand eared grebes, eleven thousand least sandpipers, nine thousand avocets, and twenty-seven thousand California gulls.

The Los Angeles Department of Water and Power had recently announced that it wanted to use groundwater for the shallow flooding. Phill, along with the local Paiute tribes, was concerned that this would cause the lake's remaining natural springs to go dry. LADWP, meanwhile, claimed that using groundwater was the only cost-effective way they could keep up the flooding. "Like they don't realize they're obligated to do it!" Phill said with a jocular indignance. He'd written them multiple stern letters of concern. "I call them my love letters."

Phill pointed out some baby cottonwood and willow trees that had established themselves on the lakebed, with help from the water from the springs. No one ever expected trees to grow on the lakebed, he said, because early in the mitigation process, "woody plant tests" had been conducted, and none of the species could survive the dust storms. But once the dust was controlled, the trees simply appeared, establishing themselves as volunteers.

Many of the saplings were sprouting from gravel, a dust control method LA started using in 2006. It wasn't used anymore, Phill said, because the state decided that it didn't look natural enough to comply with the public trust doctrine. Ironically, though, the gravel blankets had become more natural-seeming over time, because sand had blown into the rocks and plants had established themselves. It was harder and harder to tell where the gravel ended.

"When people ask what the future of Owens Lake is going to be, it's probably going to be like this," he said, gesturing at the blurry boundary between the shrubland at the edge of the lake and the gravel cover. "We're going to have fewer hard lines, and it's going to look less engineered. We've learned that the things you see in nature work the best."

Gravel getting swallowed by the lake soil, springs still spouting water into an empty vessel, baby trees. Beyond anyone's expectations, the dust mitigation efforts had made an ecology whose importance extended far beyond controlling dust.

AWAKE AT 2 A.M. the night before we flew to the US, I checked and rechecked my three piles of paperwork. There was one for our apartment, one for Lázaro to travel in cargo on the plane, and one for Mariana's immigration process. It was too much to keep track of, and I was terrified that a misspelling or oversight would hold us up at the border and ruin everything.

Sure enough, there was an error. Lázaro's veterinarian hadn't included his phone number on the letter he signed for us, even though the AeroMéxico guidelines we'd printed out for him had listed it as a requirement. I lay awake, shivering in worry, until our alarm went off at 4:30 and we headed for the airport, Lázaro's crate barely fitting into the back of the UberXL. As we waited in the slow-moving customer service line, I wanted to cry again.

"Watch," an older man, an employee of the airline, said when he saw my anguished face. He carefully leaned a husky in a crate onto a dolly. "Just like this. Everything will be fine." I smiled against my will, grateful for this man whose job it was to be gentle

                                SALT LAKES

with animals. When we finally advanced to the counter, I laid out all the paperwork, waiting for the employee to catch the mistake. My heart pounded and my hands shook. The desk agent gathered the documents into a pile with a quick sweep of her hands, not noticing the missing phone number, and handed them back to me. I slithered away feeling like I'd robbed a bank.

We'd made it onto the plane. All we had to do now, I thought, was get through customs. When we touched down, I pulled out my phone to call my mom. A Gmail banner notification popped onto the screen reading K1 INSTRUCTIONS LETTER. During the three hours of the flight, we'd been assigned an interview date. September 14, in Mexico City. Three and a half weeks away. My brain started to spark like I'd had too much coffee.

Getting interviewed meant that, if all went well, we'd re-enter the United States using the new visa, get married, and submit the green card paperwork. Then, we'd be obligated to stay in the country until Mariana's green card was processed, which could take up to a year. We hadn't prepared to come to stay. Suddenly, our lives had changed.

The checklist of documents that we needed for the interview seemed longer than before. I spent our week in Denver frantically trying to track them all down. I ordered my birth certificate and photocopied our passports and Mariana's US tourist visa. I printed my tax returns as evidence that Mariana wouldn't become a ward of the state and photos as evidence that the relationship was not only real but had deepened since the initial visa application. I booked our flights back to Mexico. On the State Department website, I paid a bot named Yatri $265 for a biometrics appointment. I called the one embassy-approved doctor in Mexico City to schedule the required medical exam, which cost ten times the price of an ordinary physical in Mexico. We had to pay extra for vaccines, because Mariana's records had been kept as a handwritten list by her mother on a piece of paper torn from a spiral-bound notebook, which we knew the government was not going to accept.

The document whose sourcing most mystified me was a certificate attesting that Mariana had no criminal record. After some searching, I found a portal on the police department website and uploaded a scan of her voter ID. They sent us the certificate as a PDF the following morning. That was easy, we said to one another with surprise and relief.

When we arrived at the college, my birth certificate was waiting for us. One more thing off the list. I covered the plastic folding table in our living room with the paperwork, checking obsessively that we had everything. When I lay in bed at night, I revolved through the list of documents just as I had before we left Mexico, mentally checking everything and worrying about everything that could go wrong. On one of those nights, a week before the interview, the image of Mariana's police certificate snapped into my mind's eye. I could visualize it down to the green logo of the Mexico City government.

The green logo of the Mexico City government. I jumped out of bed. We'd gotten the certificate from the Mexico City police, not the federal database. I went to my office and turned on the light, wondering if any of my students were awake and would notice the light from their dorm across the lawn. Sitting cross-legged in my office's armchair, I searched for instructions for obtaining a federal certificate. It had to be done in person, with an appointment made ahead of time, I learned. I went to the dining room table and found Mariana's passport in the dark. I uploaded a scan to request an appointment for the Friday before the interview.

We flew back to Mexico and shuffled dutifully between stations—the round-the-block line at the Decentralized Administrative Body for Prevention and Social Readaptation to get the criminal record certificate, biometrics at six the following morning, medical exam across the street at seven. When the last document we needed came to us as a PDF the night before the interview, we rushed to a copy shop at 9 p.m. to print the image.

The following morning, we put on button-down shirts and formal shoes and walked to the embassy. The guards told me to wait in the street in case the officials needed to question me as well, as the seating areas inside were closed due to Covid restrictions. They took Mariana to a row of kiosk windows, where she handed over the folder of documents we'd prepared. The consular official behind her window looked through them, then held a black-and-white image of a face to the safety glass separating him from Mariana.

"Who is this?" he asked.

Mariana tried to find a face, and gave up.

"How about this?" he asked, holding up another pixelated face. She shook her head. He prompted her, curtly, "They wrote your affidavit letters."

Now Mariana understood. When my sister and a friend had written letters affirming that our relationship was real, they'd had to include scans of their passports. The consulate had enlarged the two-by-two photos to fit a piece of printer paper, rendering the years-old photographs of the two women in low-resolution grayscale and at a bizarre aspect ratio.

Once she identified them, the officer sent her to a small room with a different consular official for her interview. He started off by asking how we had met. Then, he asked her why I was in Mexico. Mariana told him it was for my doctorate. When he asked the discipline of my PhD, she froze.

"I don't know," she told him.

The official smiled.

"*¿Usted preferiría hablar en español?*" he asked, eager to use his fluent, gringo-accented Spanish. She would, she said.

He asked her if she had any tattoos.

"None?" he confirmed, as though doubting her.

Outside on the street, I watched another lesbian couple celebrate passing the interview. I paced the cordoned-off block, trying to get my heart to pump more slowly. Finally, Mariana came out and said that they'd approved her visa.

"I wished I had a Virgen de Guadalupe covering my whole back," she said when she told me about the tattoo question and the otherwise friendly officer's incredulousness. "So that I could have pulled up my shirt and said, 'Oh, I forgot. Just this one!'"

PHILL STARTED WORKING at Great Basin Air Quality Control District as a data analyst in 2005, in the midst of the agency's second fight with Los Angeles. "I feel like I was born into war," he said. The two organizations reached that fight's settlement the following year. It required Los Angeles to build another thirteen square miles of dust control, and also included provisions for wildlife, water conservation, dust management, cultural resources—meaning that tribes were involved for the first time—and public engagement. LADWP hired a landscape architecture firm to build a visitors' pavilion on the lakebed. Surrounding it, they built fourteen landforms out of local dolomite gravel that they called Whitecaps, inspired by imagining wave action on the historic lake. They were meant to create habitat for insects, reptiles, and small mammals. They also built islands in the shallow flooding areas whose surfaces were tilled into curving lines. Feeling playful, they carved the words "tweet tweet" into one of them,

SALT LAKES

as though to attract birds. On another, they etched the name of their firm.

"It was the world's only advertisement visible from space!" said Phill. "We all thought it was funny, but LA did *not*."

We pulled up to the parking lot of the visitors' area, a concrete plaza with metal shade structures. There was a car in the parking lot, which left as we pulled up.

"I hope we didn't scare them away," Phill said. "That's probably only the second person I've seen here the whole time I've worked here. Once, I saw a woman painting, and she told me I was the first human she'd seen in five days. People don't know they can come here. It's kind of sad, because I think this is a really cool place. Wouldn't you come here?"

"For sure," I agreed. I hadn't told him that we'd gotten married at the lakebed. It had seemed easier not to get into it. Now, I regretted having been so reserved.

WE FLEW TO CALIFORNIA again with a deadline to get married and submit more paperwork. We summoned a couple of friends to the desert to serve as officiant and witness. The morning of our wedding, I put on a blue shirt from downtown Denver's Rockmount Ranch Wear store that my dad had given me for Christmas years before but that had felt too formal to ever wear. I tucked it into stiff, expensive A.P.C. jeans that I'd likewise owned for years and rarely worn. Mariana wore a straw hat and a red gingham shirt, and my friend Julia, a denim shirt and beaded belt. Ernesto took the cowboy dress code the most seriously of anyone: he dressed in all black, from his leather boots to his Stetson, except for the white piping and charro-style embroidery of his snap shirt.

We squeezed ourselves into Ernesto's car, drove out the college's long, cottonwood-lined driveway, and turned onto the two-lane highway that ran through its valley's expanse of sage-brush, winterfat, and bunchgrass. We crossed a mountain pass and headed south on Highway 395, down the Owens Valley. At the county courthouse in Independence, we applied for our marriage license on a glass case displaying minerals from around the county: molybdenite from Union Carbide Corporation's Bishop Creek Operation, kaolinite clay from the Black Springs Clay Mine. Outside, two separate groups of tourists asked for photos with Ernesto in his cowboy–priest getup. "I thought, I won't ruin it for them by telling them I'm a tourist too—not just a tourist, but a *Mexican immigrant!*" he told us when we returned.

Then we headed south again. I didn't exactly know how to get to the lake. I kept my eye out for possible inroads, and after passing some abandoned metal silos, I saw one. We crossed the oncoming traffic and followed a dirt road until it ended at a pair of large concrete structures. Stepping back, we realized they were the remnants of a former railroad bridge.

We set up a picnic in a wide depression in the ground, shaded by a cottonwood tree and surrounded by burrobush. We unpacked cherries, sandwiches, and tortilla chips. Ernesto pulled out a bottle of Tío Pepe sherry. It was a surprise: he'd consulted Mariana's sisters. Their dad had bought a case of it when each of his daughters was born to save for their weddings, but Mariana's had spoiled. When we first met, she had told me it was a sign that she was destined to never marry. But there we were, less than two years later, sitting in camp chairs with our friends around a wool blanket with yet another folder of paperwork at our side.

"*Ya se rompió el día,*" Mariana said: it's past noon, we can drink.

After the picnic, we headed for the shimmering rectangles in the distance. They were the shallow flooding ponds built as part of the first settlement, though I didn't know it at the time. We turned off the highway at a sign labeled Sulfate Road. On the left-hand side stood a group of low-slung buildings surrounded by a parking lot full of white trucks labeled "Los Angeles Department of Water and Power." How eerie to see the presence of the city so far away, our friends remarked.

We parked and began to explore the rectangular pools. They were each several car-lengths wide, separated from one another by earthen berms, and long enough that we didn't attempt to walk to the opposite end. Some were filled with water; others had only a white and red salt crust decorated with ridges. Moist salt clumps stuck to our shoes, like they had at the Bonneville Salt Flats. Finally, we found a pool whose clear water reflected back the enormous sky, offering a perfect reflection of the mountains.

Mariana, Ernesto, and I gathered on the salt beach at the water's edge. Ernesto hadn't prepared anything to say and Mariana was too shy to kiss in front of anyone, even just a few friends. Lázaro ran figure eights around and between us. Julia took photos. We blew bubbles from bubble wands she'd gotten at a dollar store on the drive.

"I pronounce you married!" Ernesto said, and we laughed. That was it? We went to the saloon in the nearby town of Lone Pine, ate a chocolate cake that Julia had baked and brought all the way from Berkeley, played Mexicans vs. Americans shuffleboard, and eventually walked down the block to the town's taco truck for a late-night dinner, chatting with a trucker who was ferrying apples from Washington State to Tijuana.

THOUGH THE SHALLOW FLOODING cells had had unexpected ecological importance, and though the second settlement required the LADWP to expand the area of dust mitigation, the agency didn't want to build more of them, unwilling to use more of their valuable water. They started to experiment with alternatives that didn't involve water. One of them was gravel. Another was "managed vegetation," where they planted saltgrass like a row crop. That idea, Phill said, came from Nabor Carrillo's efforts at the Texcoco lakebed. "The farm!" he exclaimed as we approached the rows of plants, tufts of green sprouting from the gray lakebed, each with crystals of salt clumped around its crown. They also added "moat and row"—fences and furrows meant to slow wind speeds and stave off big dust storms—but the state lands commission rejected it, disliking how gridded it looked.

Toward the center of the lakebed, on areas where a salt crust could form, they created laser-leveled cells to which they added brine. Most of it had hardened into a crust, but certain cells held deep, yellow-brown water that the wind was blowing into dramatic whitecaps. "I *love* brine," Phill said.

Last, they invented tillage—rows of specially-formed clods of dirt whose rough surface area kept them from disintegrating. To get the design right, they had to form a "clod subcommittee."

"We had a meeting where LADWP said a clod was the size of a golf ball; we said it was the size of a baseball," he said. "Then someone held up their hand and said, 'It's this size,' and I had to jump in, like, 'No throwing fists here!'"

They never completely figured it out. Though the clods had held together so far, Phill wasn't convinced they were failsafe. The clod areas had pumps that could quickly flood them as a backup. "I wouldn't say I don't like tillage, but . . . tillage is what keeps me up at night," he said.

In 2014, LADWP and the air quality district reached a third and final settlement. After that settlement, Phill's predecessor as agency director, who had led all three rounds of litigation, finally stepped down. Phill became the boss.

The third settlement required dust control on an additional fourteen square miles, bringing the total mitigation area to its current size, nearly fifty square miles. It will likely remain that size in perpetuity, because the agreement came with a major caveat: the air quality district agreed not to sue for any more square mileage. They agreed on a "regulatory shoreline," a literal and metaphorical line in the sand demarcating the area Phill's agency was allowed to regulate, and beyond which LA had no compliance obligations. Because of the regulatory shoreline, Great Basin wasn't allowed to hold Los Angeles accountable for dust from the dunes that surrounded the lake, for instance, which had become the area's main source of emissions. Still, at the time it had felt like an acceptable trade-off: both sides were tired of litigation.

"After that, we buried the hatchet," Phill said.

"That makes sense," I said.

"Yeah, I hope that machinery isn't disturbing it," he said, slowing down to watch some workers excavating near one of the shallow flooding ponds. "The LA guys turn over so often that they might not know it's there."

"Wait," I said, "you buried a real hatchet?"

"Of course," he said matter-of-factly. "We had a ceremony. We buried a bunch of court documents and a DVD too—it's all right over there." He pointed, as though that were how all government agencies settled their disputes.

Looking at our wedding photos, you can imagine that the line of mountains and water extends forever—the beautiful open landscape that the lake once was. They don't show the mess of pumps and pipes popping up just outside the frame, the scattered orange-and-white utility markers, the boxy air sensors, or the warning signage. They also don't show the paperwork, the payments, the fear and anxiety of waiting for decisions about your life to come in the mail. The frame cropped out the bureaucracy.

Looking back, too, I can see that it was always going to work out. It was a paint-by-numbers process; all I had to do was follow the instructions. The simple fact of having a migration pathway open to us was a privilege: many people have no recourse at all. On top of that, I was educated and adept at navigating the system, as confusing as it was. "The advantages of our whiteness and our class are clear to us at every stage," writes Ro Skelton in the essay "Little Starts," about their own lesbian green-card marriage. When Mariana's green card came in the mail, it didn't feel like an accomplishment. It felt like something no one should have to do.

Similarly, no one thinks what has happened at Owens is a good path for other lakes to follow. It is too complex and expensive an outcome. By 2024, LADWP had spent $2.5 billion on its mitigation efforts at the lake—and those expenditures will never end. The white trucks with the seal of the City of Los Angeles driving up and down Highway 395 will be a fixture of valley life forever.

Nevertheless, as climate change and human demand make less water available to salt lakes, and as politicians stall efforts to conserve them, more and more may look like Owens Lake—intricate geographies of dust control and managed habitats, including smaller and shallower bodies of water built under regulatory directives. Lake Texcoco is one example of this, mixing small bodies of water with dust-control tactics. It's also what's happening at the Salton Sea, where the California water board has mandated the construction of 15,000 acres of habitat and dust suppression by 2028, starting with shallow "habitat cells" at the south end of the sea and a wetland at the north end. At Great Salt Lake, 512,000 acres of lakebed—eight times the area of Owens Lake— have already been exposed. Some believe that a lawsuit there based in the Clean Air Act may not be far off.

Whether or not it is a lake, and whether or not it is a good outcome, Owens is another queer ecology. Improbably, through

                                              SALT LAKES

persistence and experimentation, life has returned. Brine shrimp live in bermed-in trapezoids, snowy plovers build nests in gravel laid down by trucks. Local birders say more birds come to Owens Lake now than did a century ago when it was a lake, because of the mix of surfaces and habitats. The engineering has facilitated wildness and fecundity rather than stripping life away.

Both my marriage and the life that has returned to Owens Lake offer proof that vitality can flourish out of the constraints of bureaucracy. My wedding—held at the beckoning of borders, ordered on a strict timeline by the authorities at United States Citizenship and Immigration Services, painstakingly documented and still subject to revision and verification—didn't feel like a "real wedding," inasmuch as it didn't resemble the events held by any of my friends or family. But it had its own emergent properties. It made two girls who giddily tell each other that they look cute in their raincoats, who spend long mornings in bed with their dog, who refer to one another as "*lombriz*" and "*sarigüeya*" like a lesbian, Hispanophone *Frog and Toad*, into each other's legal family. The bureaucracy allowed us to cleave ourselves from the families and nationalities we had come from and replace them with our own chosen, horizontal tie.

The novelist Garth Greenwell writes in an essay, "it seems plausible that gay people have changed marriage as much as it has changed us." There's a queer pleasure in misrecognizing marriage—pretending to play by its rules while in fact taking one's newfound rights and benefits to live in a way the state has no capacity to imagine. Queers have opened up what counts as marriage; we've expanded and exploded the rhythms and practices of life that its legal bonds comprise.

Similarly, though the Owens may not count as a "real" lake, it's still something real, existing in the world in its strange and unique way, providing habitat to plants and flies and birds. It has opened up what counts as a lake. If it provides enough water to keep the lakes' food webs intact, it seems to me to be a compromise we can live with. In choosing to see the Clean Air Act–mandated landscape of dust mitigation as habitat, that queer pleasure of celebrating the uncanny as ordinary is present, too. The lake comes back to life. Queer ecologies flourish in the interstices, and the future is full of such gridded, broken, unruly places.

CHAPTER 8

WILLING SELLERS AND THE PUBLIC TRUST

Walker Lake, Nevada, USA

WE PACKED OUR VAN AND HEADED NORTH FROM Tucson, the border city that had become our US home. We passed Las Vegas as it was getting dark and slept behind a Chevron truck stop in Indian Springs. The light rain falling as we pulled out of the massive parking lot the next morning turned, further north, to snow and hail. The snow made the landscape look soft, like blankets with rocks sticking up out of them. It's a soft geology: Nevada's mountains are different from the Sierra Nevada or the Rockies because they don't have foothills; they slope gently all the way to the basin floors.

On one mountain pass, I weaved around two semis stuck diagonally across both lanes of traffic. On another, we waited in white-out conditions behind several stopped trailers in fast-accumulating snow before I snuck between them, seeing only a few feet ahead, sure that if we remained any longer we'd be snowed in. When we finally reached the gas station at the south end of Tonopah, it felt like safe harbor: we'd passed the storm. We headed west, the snowy landscape now bright under blue skies, until the turnoff for our destination, a small town squished between two Department of Defense installations. It announced itself with a vertical sign spelling HAWTHORNE in lights, as though the town were a vintage movie theater.

I was there to meet Glenn and Marlene Bunch, a retired county maintenance worker and travel agent in their eighties who have been fighting for over thirty years to restore Walker Lake. Those who have heard of Walker Lake know it as a place where Las Vegas gangsters would dump cars in the 1930s. It's unusual among endorheic lakes because, until recently, it was one of only a handful in the world fresh enough to support a fishery and freshwater birds. But the Bunches had seen it dry up. Since the start of irrigation in its basin, the lake has lost more than 90 percent of its volume.

The Bunches lived in a modest blue-gray house on a corner lot. It had a large pine tree in front, along with a Jeep, an F-350, and a DMC 1450 mini Snowcat, all labeled "Mineral County Search & Rescue." In addition to the Walker Lake Working Group, Glenn sat on the board of the Nevada Department of Wildlife and was the commander of the county search and rescue; both are also active in their church. As she welcomed me into the house, Marlene, who wore a necklace with a stone in the shape of Nevada, joked that everyone in the county deals with their problems with 1-800-CALL-GLENN.

Marlene called the dining room "command central." It had two desks with hutches stuffed with binders and radios. One of them had two computer monitors on it, each with a different Gmail account open—one for the Walker Lake Working Group, the other for Marlene's church group. At the back corner of the desk, she had a store of extra USB drives in a painted-glass Garfield mug, the kind that came with Happy Meals in the 1980s.

Sitting on the desk, too, was Glenn's gauge log. Each morning between his first and second cup of coffee, Glenn sat at his computer, opened up the Walker Basin Hydro Mapper—a platform developed by the US Geological Survey to visualize the flow gauges along California and Nevada's Walker River—and wrote down the gauges' readouts on a 5x8 legal pad that he had divided into five columns with thick, ruler-straight pencil lines: Date – Flow – Elevation – Volume – %.

On the wall, next to the thermostat, hung a small whiteboard. It had a to-do list with tasks like "trim trees" and "wiper blades." The top item read "3,950." That's the minimum goal for the lake elevation, Marlene explained—a water level that would allow it to be stocked with fish again. They hope to reach that water level by the mid-2030s. The day I visited, the lake was at 3,907.68—still a ways off. "I've asked God for eleven more years," she said.

THE WALKER RIVER BEGINS in California's Sierra Nevada above the town of Bridgeport, just north of Mono Lake. From there, it flows north and east through Nevada's Smith Valley. Along its banks grow native plants like willow and buffaloberry, overlooked by dramatic cottonwood galleries. Further away, on the valley floors, are salt-tolerant shrubs like rabbitbush and species of *Atriplex*. After the river passes the town of Yerington and its former copper mines—now Superfund sites—it continues its sweep northeast, circumventing a mountain range before flowing south through the Walker River Paiute Reservation and its main town, Shurz. From there, it fans out in a delta into Walker Lake. But for much of the last century, it never got there.

In 1859, Congress created the Pyramid Lake Paiute and Walker River Paiute reservations. They were sister reservations, each with a lake, as John McMasters, the Walker Lake Paiute Tribe's Director of Land and Water Projects, told me. The tribe's traditional name means "trout-eaters," referring to the native Lahontan cutthroat trout. But the settler government didn't want the tribe to continue living by their traditional means and imposed a cultural transformation to agriculture and ranching, educating the tribe to use water from the Walker River for irrigation. At the time, there was no state water law, nor was there anyone competing for the water.

In the late 1800s, after the Homestead Act and Desert Lands Act, white settlers started arriving. In response, the federal government removed Walker Lake from the reservation boundaries, putting it, and its adjacent lands, back into the public domain. They did so because they thought there might be gold in the mountains surrounding the lake and a settler might want to make a mining claim.

As settler irrigation increased, the river ran weaker and weaker. In 1902, California cattle baron Henry Miller brought a lawsuit over the Walker River's waters. He wanted to prevent a rancher upstream from diverting the river into a reservoir, thereby decreasing flow through Miller's land. The dispute went to the US Supreme Court, which determined in 1910 that since the Walker Basin spanned two states, a federal court should oversee the river. They assigned the case, and the management of the river's basin, to the United States District Court in Reno. (The district court continues to oversee the governance of the river, and locals refer to it as the "decree court." The "federal water master"—a term harkening back to Brigham Young's system of allocating water in the early days of Mormon settlement in the Salt Lake Valley—has an office on Yerington's Main Street.)

In 1919, the court issued a decree about the allocation of the river's water, earning it the "decree court" nickname. It didn't include provisions for the tribe, however. So, in 1924, the United States, operating on behalf of the Walker River Paiute Tribe, brought another lawsuit. The decree was rewritten, giving the tribe the most senior water rights. But they were rights for irrigation only, not for sustaining traditional foodways like the fishery.

"I think it's a blatant violation of Indian trust," Michael Blumm, a professor of environmental law at Lewis and Clark Law School, told me. "If the federal government had recognized that one purpose of the reservation was to sustain a principal source of food—fish—they would have asked the decree court for water for the fish. The Justice Department just screwed up, but they did that a lot back then. They thought of water as for growing crops, which is disastrous for users like the tribe."

The 1936 settlement was supposed to be final. But, over time, the tribe realized that they weren't happy with it. They had not been allotted enough water for their needs, and not even all the water to which they were entitled seemed to reach their small reservoir. That was because the river had been over-allocated upstream, allowing the hundreds of farmers that formed the Walker River Irrigation District, or WRID—which is headquartered in Yerington in the same building as the water master—to use up all its flow before it reached the reservation.

So, in 1993, the tribe sued WRID. They made claims to water rights for their reservoir and for lands that the government had restored to the tribe since the 1936 decree. Later, they added a claim to rights for groundwater.

South of the lake, in Hawthorne, the newly formed Walker Lake Working Group came up with an idea. If the decree could be reopened, could the lake itself also demand water rights? In 1994, the group persuaded Mineral County to join the case. Inspired by the recent success at Mono Lake, they asked for a minimum of 127,000 acre-feet of water per year to be provided to the lake under the public trust doctrine.

WHEN GLENN AND MARLENE were raising their kids, Walker Lake was a place for fishing, boat races, and water-skiing. Hawthorne residents spent the summer camping on its shores. The Bunches had a boat, too—and a parachute. It was the only thing big enough, explained Glenn, to provide shade for all the friends they invited out to the beach on the weekends to hang out and water-ski.

"People would be driving along and see the parachute and say, 'Oh, the Bunches are there,' and just come join us," Glenn said as we ate the breakfast Marlene had prepared, a baked casserole with layers of sausage patties, hash browns, scrambled eggs, and cheese.

Still, they knew the lake's water level was dropping and its salinity increasing. Its Lahontan cutthroat trout had disappeared in the 1950s. After that, the Nevada Department of Fish and Game had stocked the lake with hatchery-raised trout.

In the late 1980s, a biologist with the Nevada Division of Wildlife obtained data about the decline of the Lahontan cutthroat trout and tui chub in nearby Pyramid Lake. The information had been sealed for years as part of a court battle. Using the Pyramid data to hypothesize about Walker, he realized that the lake was just a few years away from losing its fish.

The Bunches remembered that history in scene. As they told it, that year they volunteered at a fishing derby that the Nevada Department of Wildlife hosted to collect data about the lake. As Glenn was launching the boat, he and Marlene chatted with the biologist. "The lake had just been going down and down," Marlene recalled. "So I asked him, 'How long before we don't have a lake anymore?'"

"He says, 'It'll never dry up,'" Glenn said, picking up the story. "'It will recess down to where it's so thick that it's just a sludge pond, but it'll never dry up, with all the dissolved solids. It'll be stinky. That's where it's going to end up.'"

Marlene asked when they would lose the fishery. The biologist told them that trout wouldn't survive once salt concentrations reached 14,000 milligrams of dissolved solids per liter, and that they'd reach that threshold in the early 2000s.

They did, right on time. Each year, the hatchery biologists would come with a live cage—a net they would put into the lake to monitor fifty fish for twenty-four hours.

"It started out we'd lose five or six," said Glenn. "Then we'd lose ten. The last time we put them in, we lost 100 percent in twenty-four hours." At that point, the hatchery told them, they couldn't waste any more state money on stocking the lake.

By then, the Bunches and other Hawthorne residents had formed the Walker Lake Working Group. The group caught the ear of Senator Harry Reid, who had made his love for Walker Lake known during past campaigns. With funds he secured, the working group hired a lawyer named Simeon Herskovits. He was a former New York City corporate lawyer and frequenter of punk shows at the famed CBGB club who had decided to change his life and specialize in water issues in the rural Great Basin.

With the Mono Lake decision fresh, public trust was "pulsing through the public interest law community," Kyle Roerink, director of the Great Basin Water Network, told me. Herskovits wanted to apply it at Walker. "It was a quintessential David vs. Goliath fight," Roerink continued. "He told them there was this concept of public trust that had been validated in another legal jurisdiction and that would be possible to get recognized there, too. And that it would be a means of affirming their rights as citizens and holding the state accountable to its foundational common-law principles."

For years, Walker Lake's public trust case couldn't move forward because of a logistical problem: serving papers to all the defendants. When the Audubon Society and the Mono Lake Committee pursued a public trust argument for Mono Lake, they were taking on just one foe: the Los Angeles Department of Water and Power. But at Walker, the federal judge required all 1,200 water users of the Walker River Irrigation District to be notified of the case.

"The weekend we were going to go serve the papers, they put the word out in the newspaper," as Marlene told the story. "All the big ranchers went over to their California ranches for the weekend. And they said that absolutely nobody, none of the hired hands or anybody, was to surface if we showed up." The effort to serve papers to the defendants ended up lasting years.

In the meantime, Senator Reid kept finding funds for the lake. As part of the 2002 Farm Bill, Congress approved a transfer of $200 million to the Bureau of Reclamation to create the Desert Terminal Lakes Program. The money would go to a combination of purchasing water rights for the lake and conducting scientific research about its watershed that would support those transfers.

The National Fish and Wildlife Foundation was put in charge of purchasing water rights, while the University of Nevada took the lead on the research. The combination made the Walker Basin—a place few people had ever heard of—into one of the most studied watersheds in the nation, and also put the basin at the cutting edge of technological monitoring of water diversions. Most watersheds don't keep data about how much water is diverted at each point. But in the Walker Basin, the USGS installed a system of gauges that monitored the entire length of the Walker River and projected their readings online on the Hydromapper, allowing scientists and the general public alike to see where water was going, and how much.

Many farmers in the area didn't take kindly to the federal government showing up and offering to buy their water. Though the National Fish and Wildlife FODedation—"Nifwif," as it's colloquially known—emphasized that it was there only to buy from willing sellers, not to coerce anyone into selling their water or land, their presence was perceived as an existential threat. "As there is through much of the West, there was a fear that the water was eventually going to be taken," explained Brian Masini, a farmer who was one of the first to sell water rights to the program.

The state was also suspicious. For each water right that the foundation purchased, it had to submit a change application to the Nevada state engineer, who would convert the water's "manner of use" from irrigation to in-stream flows. Until the application was approved, the water had to continue to be diverted—or, in water terms, put to "beneficial use." If not, the right would be considered wasted and the foundation would risk losing it.

The state engineer's oversight dates back to the early days of water rights and the Colorado doctrine. Lawmakers and judges made it difficult to transfer water rights in order to prevent water from accumulating into just a few hands. To receive approval, applicants have to prove that the transfer won't create negative effects on other rights-holders. But when the foundation submitted its first change application in 2012, the state engineer received over forty protests, stalling the process. "We're ruled by the 1800s here," one person told me with a resigned laugh. A legal order finally moved the application to the decree court for approval, where another comment period attracted complaints from far beyond the irrigation district. The water didn't reach the lake until 2019.

As years passed, things got worse. Dave Herbst—"Bug," the insect biologist who had been part of the baseline survey of Mono Lake—started collecting data about the lake's invertebrates. By then he was a professor at the University of California–Santa

Barbara. He had begun visiting the lake in the 1980s. But when he returned in 2007, it was a changed place. The lake's salinity was 19 grams per liter, up from its historic 2.5. Though the trout were gone, some native tui chub fish were still hanging on, and many waterbirds still stopped at the lake.

Using nets and buckets, Herbst and others collected larvae and pupae specimens from four different types of lakebed: sand, gravel, cobble, and an algae–widgeongrass mix. They found that the lake was home to four invertebrates: two types of midges, one type of damselfly, and one worm from the genus *Monopylephorus*. These insects didn't live at saltier lakes, and their population makeup showed that things had changed at Walker. The *Hyalella*, a crustacean that resembles a translucent shrimp, had been the most abundant invertebrate during Herbst's previous trips, decades earlier. Now, it had disappeared completely.

Increasing salinity was not only affecting the invertebrates directly, it was also changing their habitat. Salt made it harder for widgeongrass to grow, and diminishing lake levels left the rockier areas encircling the lake—which the creatures preferred to the muddy basin at its center—exposed.

Herbst came back again in 2012 to conduct an experiment similar to the one he had conducted during the Mono ecological study, testing the stamina of the lakes' bugs. He and his collaborators created tubs of water at a range of salt concentrations: 10, 15, 20, 25, 30, 40, 50, and 75 grams per liter. To keep the salts consistent, they used the lake water itself, either evaporating it to achieve higher concentrations or adding distilled water to dilute it. Then, they harvested 400 of each species of insect and placed each one in a 50-milliliter container of water. After 12, 24, 48, 72, and 120 hours, they tabulated the insects' mortality. The damselflies' mortality increased steadily with increasing time and salinity; the midges died off in large groups after certain thresholds.

They also measured the size of the remaining living insects. As they had observed among the flies and shrimp at Mono, for the bugs of Walker Lake, trying to survive at higher salinities came with a major trade-off. Those that managed were significantly smaller, in both length and weight, than their counterparts in less saline water; they'd had to expend their growing energy on cleaning their body of salts. "The survival of the benthic invertebrates of Walker Lake is near a limit," he wrote. He was right; soon after his study, the insect populations collapsed and the tui chub disappeared.

As at other lakes, the decline of the invertebrates resounded through the food chain. The midges and worms fed the damselflies, which in turn fed the tui chub, which fed the birds. Walker Lake had been famous as a stopover point for loons migrating to Canada. But when the insects and fish died, "they couldn't refuel," as Glenn put it. "There was nothing to eat." They died on the shore, as the eared grebes at Great Salt Lake would later. Then, they stopped coming.

"We've had so many heartbreaks over this," said Marlene.

SLOWLY, things fell into place. In 2014, while the first change application was still in process, the National Fish and Wildlife Foundation handed over the willing-sellers program to the newly created Walker Basin Conservancy. By 2025, after more than twenty water rights purchases, the conservancy had obtained 59 percent of the rights necessary for their goal of sending 50,000 acre-feet of water to the lake per year. The state had also become more cooperative, making it possible to complete each change application in under a year.

I visited the conservancy's office in early March 2023. The year's irrigation season, which runs through the end of the summer, had begun just two days earlier. The conservancy's headquarters were located at the northeastern edge of Yerington, around a curve from the town's quaint, all-America-style Main Street. The parking lot of the small cinderblock building with a blue metal roof was full of white trucks; beyond it there were two greenhouses. The office manager explained to me that when the conservancy buys water rights, they tend to buy the land that comes with them. In addition to administering water, then, they had created a habitat restoration program, turning formerly irrigated cropland back into a Great Basin ecosystem. Eventually, those lands become public parks: they've opened over 13,000 acres to the public. But after realizing that a major limiting factor on their work was access to native plants, they started to propagate them themselves. They now were expanding greenhouse operations to allow for public purchase as well as internal use.

"I love the desert," conservancy director Peter Stanton—a marathon runner and backcountry skier originally from San Antonio, Texas—told me over lunch at El Alteño, a Mexican restaurant he said had the best food in Yerington. "I guess that goes without saying if you're going to run a 13,000-acre saltbush restoration program."

We got into his white conservancy truck and drove to see the basin. On the southwestern edge of town, between a complex of beige warehouses and the abandoned open-pit mine, I thought I saw shredded plastic flying through the air in front of me. Then I realized they were pieces of onionskin. Historically, most of the farmers that made up the Walker River Irrigation District grew alfalfa. In recent years, they had started planting onions and garlic, which, in theory, use less water than green forage. But many farmers made the shift because it allowed them to have crops in the ground year-round. Because of that, the row crops can end up using more water.

Leaving town, we drove to Pitchfork Ranch, a property that the conservancy had acquired through its water rights program and donated to the Nevada State Parks agency. It had opened to the public even as the conservancy continued its restoration efforts. The Walker River flowed along the edge of the park, lined by cottonwood trees and other native riparian species. On the former alfalfa fields further from the river, seedlings and shrubs were planted in contoured lines connected by drip irrigation. Stanton got excited when he saw plants outside the lines: it meant the seeds were dispersing naturally and the plants establishing themselves on their own.

Back in the truck, we followed the river downstream to the Walker River Paiute Reservation's Weber Reservoir. The reservoir used to be the end of the line for the river's water—since not enough reached it even to meet the tribe's water right, there was nothing left to flow on to Walker Lake. When the water rights purchasing program began, the tribe agreed to take responsibility for releasing water from the reservoir to Walker Lake.

My visit coincided with the first day that season that they would open the headgate. Peter and I approached the small reservoir, drove across its dam, and parked. We walked over to a small outlet in the concrete. It was gushing water.

"Wooo!" Peter let out a cowboy yell. "That's what we like to see!"

In 2013, a federal judge finally granted Mineral County's motion to join the tribe's legal cause against the irrigation district. First, though, the district court decided against the county, saying that its public trust case was invalid because Walker Lake was not part of the Walker River watershed.

Herskovits and the working group appealed that ruling, and the case went to the Ninth Circuit Court of Appeals in Pasadena. The new court held that the lake was, indeed, part of its own basin, and in May 2018 asked the Nevada Supreme Court to resolve the open question of whether the public trust doctrine was part of state law.

When the court answered in 2020, it offered only a partial answer. It determined that public trust did apply in Nevada but emphasized that it could not be used to take away existing water rights.

"We recognize the tragic decline of Walker Lake," the court wrote. "But while we are sympathetic to the plight of Walker Lake and the resulting negative impacts on wildlife, resources, and economy in Mineral County, we cannot use the public trust as a tool to uproot an entire water system, particularly where finality is firmly rooted in our statutes."

At first, the statement felt like a painful blow. It was also confusing—it seemed to simultaneously affirm the public trust doctrine and throw it out as a recourse. But then, the activists realized that the fact that the court hadn't fully answered the question meant that any possibilities for conserving water without reallocating water rights were still on the table. The state could apply the public trust to change the management of water outside the irrigation season, require farmers to make efficiency improvements to their irrigation systems, or mandate that the state create a plan to conserve the lake. They could take advantage of the USGS gauges to ensure that farmers didn't take more than they needed. Using those strategies, it would be possible to mandate an elevation for the lake similar to the one at Mono.

The Ninth Circuit agreed. In 2021, they sent the case back to the Decree Court with the instruction to consider solutions to the public trust claim that didn't require taking and reallocating water rights. If the court agrees to apply the public trust in these ways, Nevada could lead an overdue update to the way water rights are managed in the West. In particular, the Mineral County public trust case could change the meaning of the term "waste." It could come to mean what it should have meant long ago: over-irrigating land and over-allocating rivers instead of allowing a river to reach its endpoint.

What happens at Walker is a trial run for Great Salt Lake, where the amount of additional water that scientists estimate is necessary to recover a healthy elevation is not 50,000 acre-feet per year but one million. Though there's a public trust case underway in Utah, too, the state's conservative politics temper most lake advocates' optimism that the state will favor reallocating senior water rights. Instead, across the West, it seems more likely that the conservation of waterways will come from combining the two elements being tested in the Walker Basin: technical, seemingly minute legal changes and the market-based solutions of purchasing water rights and paying farmers to fallow their land.

Those changes are, if slowly, underway. "In a way that's almost been difficult to perceive, our entire water law institutions are being rejiggered," Brigham Daniels, director of the University of Utah's Great Salt Lake Project, told me. In January 2024, Daniels published a paper, "Utah's Legal Risks and the Ailing Great Salt Lake," that alerted state lawmakers to three legal tools that could be used to impose environmental protections on Great Salt Lake if the state didn't take adequate action to prevent it from drying up: the Clean Air Act, the public trust doctrine, and the Endangered Species Act.

"There have been calls for reform of water law for a long, long time," said Daniels. But where those past calls envisioned a complete reinvention of the system, he continued, the public trust isn't functioning as a strong-arm solution to force the region to start over. Instead, it's offering a beacon on the horizon of a shared common good, a goal toward which small, technical changes are gradually moving. In Utah, the state legislature has begun to make water transfers more feasible, and has legalized in-stream flow as a type of beneficial use for the first time.

"Six years ago," continued Daniels, "any water that made it to the Great Salt Lake was a failure—it wasn't used, it was wasted and gone to this stupid old lake. Now, having water in the Great Salt Lake counts as a beneficial use. The 'use it or lose it' allows it. That's the experiment, and it might be enough to really reallocate a lot of water."

With those changes in the offing, the Bunches were feeling optimistic for the first time in years.

"We've been at this for thirty-one years and it's been a long hard struggle," Marlene told me as we said goodbye. "It just seemed like it was never going to end. But we're closer now than I think we've ever been."

CHAPTER 9

AN ACT OF ATTENTION

Lake Abert, Oregon, USA, and Laguna Mar Chiquita, Córdoba, Argentina

I MAGINE YOU'RE A PHALAROPE. A FEMALE WILSON'S phalarope, or *Phalaropus tricolor*. Your tiny, fragile bird's body fits in the palm of a human hand and weighs little more than an AA battery. You have long, dark legs, a white belly and blue-gray wings that lighten into a ruddy neck. Your mostly white face has a dark cap and a black mask leading to your slim, pointed beak.

This coloration makes you unusual among birds; as a female, you're larger and more colorful than your male counterparts. You look good and you know it: female phalaropes choose their mates though aggressive conflicts with other females and also practice polyandry, or keeping multiple partners.

You're a weird exception in another way, too. You're the only species of bird that migrates the entire length of the Western Hemisphere stopping exclusively at salt lakes. Your year starts with spring nesting in the prairies surrounding Saskatchewan's Chaplin Lake. Then, you fly south to the salt lakes of the Great Basin—especially Great Salt Lake, Mono Lake, and Oregon's Lake Abert—where you chow down on flies and fly larvae to double your weight before migrating. You might be best known for the unusual way you trap your prey: when they're not at the surface, you spin in the water, shoveling it backward with your lobed feet to force the fly larvae suspended in the lake to come to the surface. Once you've bulked up and molted your coat, you fly south to saline lakes in South America for the austral summer.

But what if you showed up at one of those lakes and it wasn't there?

In 2014, 2015, 2021, and 2022, that's what happened to the phalaropes and other shorebirds that flew to Lake Abert. Abert, the largest lake in the Pacific Northwest, sits in a wedge-shaped basin in the sagebrush steppe of sparsely populated Lake County, Oregon. Around 10 percent of saline lake birds spend their summers there; it boasts even higher bird concentrations than Great Salt Lake. But between dry winters and increasing water diversions for ranching and farming, Abert keeps drying up. Each time, over the course of the summer, the water recedes to a puddle of brine, the salts harden into a white crust of triangular crystals, and the tiny brine shrimp and alkali flies die off en masse.

IN SEPTEMBER 2023, I rented a compact hatchback and headed for Abert. I drove south from Bend until pine forest gave way to a broad steppe. Though this marked the start of Oregon's arid eastern half, to my desert-trained mind it was still strikingly green. Near the playa of Summer Lake, whose salt crust was blowing off in wisps, a volcano-formed ridge loomed over me, its forest burned in the 2021 Bootleg fire. To the left, I passed the Summer Lake Migratory Waterfowl and Game Management Area, a nature preserve partially created out of ranchland once owned by the family of William Kittredge, the rancher-turned-writer whose work had helped me articulate my experience ranching.

In the small town of Paisley, I turned onto River Street. According to the map, it would lead me to Abert along the Chewaucan River. I went slowly; the asphalt was disintegrating and full of potholes, and I wanted to take in the surroundings. On the right side were fields. I saw two sandhill cranes in the mowed grass, standing upright and extending their long necks. I was happy to see them. Cranes had migrated through the ranch in New Mexico, squawking loudly as they flew overhead.

To my left, the river ran alongside the road in a deep channel. The Chewaucan, which starts on the volcanic ridge I'd just passed, provides nearly all of Abert's inflow. But as it crosses the steppe bottomlands toward the lake, ranchers divert its waters into canals to flood broad, grassy marshes where their cattle live during the winter and give birth in the spring. Those irrigation systems started in the early 1900s, but have increased in recent decades. By now, if all the water rights on the river's 255 diversion points were fully used, they would exceed both the river's flow and the volume of Lake Abert. In contrast to the Walker Basin, where water use is measured by gauges and overseen by the Walker River Irrigation District, in the Chewaucan Basin it's sorted out among neighbors. It's "highly controlled," ranchers assured me, but without a paper trail. Between the effects of climate change and these diversions, the amount of water that reaches Abert is half of what it was historically.

Before long, the road was blocked by a ranch fence. I shook an imaginary fist and silently made fun of myself for the way my politics had changed. I turned around and headed back to the rural highway. It led me through the rest of town and then east to a narrow Y. I took the sharp turn onto a familiar road: Highway 395, which, hundreds of miles south, also passes Owens and Mono lakes. Another volcanic wall, Abert Rim, overlooked the lake and me.

After a snowy winter, Abert was full. I parked on a gravel shoulder and tried to approach as slowly as possible, to observe it as thoroughly as I could. From a distance, the landscape was dominated by the brown-red of lava rock and the bright yellow-white of dormant grasses. Closer in, green tufts of sage and saltbush appeared, scratching my legs as I walked. In the distance I could see the brilliant white shoreline and the water, which was darker than the sky. I pushed my way through the bushes, more and more thickly grown the closer I got to the shore, until they came to a sudden halt. On the white beach, only a few *Chenopodium* plants pushed through.

A black ring encircled the lake. From afar, it looked like the line of seaweed that washes up on ocean shores. But up close, it was buzzing: it was flies. Further away, the strands, or former water lines, were marked with stripes of gold. I got closer. They were the empty cocoons of thousands of fly pupae. The little shells were so numerous that I felt an impulse to sweep my hand through them, to feel them fall through my fingers. It was the summertime abundance that scientists had told me about. I'd never visited the lakes at the right time to see it for myself.

And then there were the birds. At first I didn't see them. They weren't on the southern shore, where I'd parked. But as I walked northward, crunching the salt blooms beneath my shoes, passing grasses with crystals of salt clumped at their bases, jumping over hordes of flies and puddles of mud, they started to appear. First, long, skinny, three-pronged prints guided my way along the salt beach. Then I started to see smaller prints. Then, the birds themselves. One floated out in the water, with a black stripe on its wings and a black back. Another stood at the water's edge, picking up flies with its thin beak and legs so long for its small body. As I walked further, I saw more. Tiny ones ran along the beach in groups. Others bobbed on the water, far enough away from me not to be bothered. Others still flew off in groups as I approached.

I knew some of their names, but couldn't match them to the birds in front of me. Plovers, sandpipers, avocets, phalaropes.

I tried to search for photos on Google, but didn't have enough cell service. I drove to the north end of the lake, which was densely packed with birds. I wished I had a pair of binoculars.

Driving back, I saw a pickup truck with a camper shell parked on the shoulder and a man peering through a long spotting scope. I had a hunch about who it was: Ron Larson, a retired US Fish and Wildlife Service scientist whose book about the natural history of Abert was coming out that fall. It isn't every day that you run into a lake's main scientist at the lake itself, I thought. I had to say hello. I braked fast and pulled onto the gravel. I broke the ice by asking what kind of birds we were seeing. They were mainly avocets, he told me. The small ones were least sandpipers.

When I started to meet and spend time with other people who cared about salt lakes, I'd felt like my knowledge was inadequate, specifically in the realm of birds. Birds were one of the main reasons people cared about the lakes. At peak season, 1.5 million shorebirds crowded Great Salt Lake, 100,000 congregated at Mono, and 350,000 at Abert. Many saline lakes had special ecological designations because of these high concentrations of the few shorebirds that had evolved unusual adaptations to salt, like glands around their eyes that get rid of the salts they ingest in their food and thicker tongues to squeeze salt water off their prey.

But while I lacked expertise, something had drawn me to birds for a long time. As a kid, I'd had a gray sweatshirt that read, "Welcome Back Old Friend!" commemorating the re-introduction of the peregrine falcon to Colorado. It's the only item I remember asking my mom to buy me after a presentation at school; she always insisted that we not be materialistic, and on top of that didn't like us to wear clothes with words on them. But the story of the return of the birds of prey had inspired me, and perhaps she agreed because she loves animals. Wearing the sweatshirt made me feel like I had a special relationship with the majestic falcon.

I also loved the American kestrel, which I knew from the square one-cent postage stamps we used every time the price of postage went up. The bird was red and blue, colors that seemed unbelievable for a falcon. I never thought I'd see one. But then, riding shotgun in the light Toyota pickup early in my time on the New Mexico ranch, I'd spotted one on a telephone line. I could identify it right away. It really was blue and red, with cream-colored patches and black dots all over. The sighting felt improbable and magical, and the click of recognition in my brain an equally impossible satisfaction.

In both high school and college, when I saw ornithology electives offered, I signed up without thinking twice. The high school course included extracurricular outings with the teachers who were birders, mainly teachers of biology but also of English, all men of retirement age who kept teaching because they loved it. I knew one of them from a lunchtime Thoreau reading group that we called Club Walden. They were pensive and Puritan, people who modeled a simple life constructed of books and nature and who gave me the idea that I could make the same. The fact that they found value in birding made me sure it must hold something profound.

The adaptations that birds have developed to be able to live at salt lakes are fast becoming liabilities rather than advantages. Since the lakes offer their birds just a few stopover points in their yearly migrations, the species are vulnerable to change. Losing any single salt lake likely means a significant population decline. With salt lakes across the world drying up—including not only the large perennial lakes, but smaller seasonal and ephemeral lakes—the future for the birds could be bleak. Already, in the Great Basin, waterbird populations have declined by 70 percent since 1973.

This also means that protecting the birds could be the key to saving the lakes. In 2019, scientist Ryan Carle realized this. Carle was born in 1985 and raised in Lee Vining, California, on the shores of Mono Lake. His parents were California state park rangers who served as the lake's resident naturalists. He grew up to be an ornithologist, studying waterbirds in the US and Chile. But after years working along seacoasts, he started wondering about his inland home. A lifelong kayaker, he had always admired the way the small phalarope lit off the surface of Mono Lake in large, fluttering groups. No one had studied the bird in thirty years; no one knew how the environmental protections at Mono or climate change had affected their population numbers.

Carle contacted Margaret Rubega, who had conducted the most recent studies at Mono Lake, during the early 1990s. Now in her sixties, a professor at the University of Connecticut and the Connecticut state ornithologist, Rubega had been the one to explain phalaropes' unusual feeding style—how they spin to bring their prey to the surface of the water and then use their long, narrow beaks to suction it off the surface, opening their beaks to catch it in one quick chomp once it's in the air.

They decided to join forces. Rubega proposed holding an international meeting of all the scientists studying and monitoring phalaropes. She and Carle hosted it at Mono Lake in June 2019. It was a relatively small group of people, and they quickly organized their priorities. They needed to increase research about the bird, taking advantage of new tagging and tracking technologies. They needed to restart population counts—the Utah Division of Wildlife had stopped conducting phalarope-specific counts ten years earlier, and no organized counts were happening at Mono, Abert, or Owens lakes, or at most lakes in South America—and to coordinate them to be simultaneous across lakes, in order to be confident they weren't counting birds twice. They needed to use those simultaneous counts to develop a new population estimate: the world population estimate of 1.5 million dated to 1988.

Why had there been so little attention paid to the phalarope? "It's such a small bird, it's often far away, and it's gray and white, so it doesn't attract a lot of attention," Marcela Castellino told me.

Castellino, a petite woman with dirty-blonde hair, is like Carle's mirror image, a long-lost twin from the opposite end of the phalarope migration pathway. She was also born in 1985 and also grew up on the shores of a salt lake—in her case, Argentina's Laguna Mar Chiquita, where more than half of Wilson's phalaropes spend the austral summer after their 3,500-mile migration from North America. She and her sisters, Mariana and Marina, are all named after the lake. She began studying phalaropes while studying biology at Argentina's National University of Córdoba, after she participated in an exchange program organized by Utah's Weber State University that brought students from Argentina, Chile, and Mexico to Great Salt Lake to develop research projects oriented around the birds that migrated between Utah and their home countries. For Castellino, that meant studying the phalarope—or, as it's known fondly in South America, the *chorlito*.

After the Mono Lake meeting, the members of the newly-christened International Phalarope Working Group got to work quickly. That summer, they initiated simultaneous phalarope surveys at the species' six most important stopover sites in North America: Lake Chaplin, Mono Lake, Lake Abert, Great Salt Lake, Owens Lake, and the San Francisco Bay Salt Ponds. They wanted coordinated counts so that they could see how the phalaropes distributed themselves across the continent's entire habitat, not just at individual lakes. For instance, when Lake Abert dried up the following summer, they saw a spike in phalaropes at Mono Lake.

At Great Salt Lake, two employees of the Utah Division of

Wildlife, John Luft and John Neil, restarted phalarope counts by plane. They flew counter-clockwise around the lake, one John looking out the left window and the other out the right. They noticed that the way the phalaropes congregated at the lake had changed since they had stopped doing surveys a decade before: previously, large flocks had stationed themselves in Great Salt Lake's shallow and relatively fresh bays. Now, those bays were dry. Instead, Luft said, the birds formed a "ribbon around the shoreline."

Working group members also began to fit phalaropes with nanotags, lightweight trackers that allowed them to trace the flight paths of individual birds. They attached them to the birds by sliding their tiny legs through two loops of fishing line—"like a pair of underwear," in Carle's words—so that the tag rode on the bird's rump. To fit the phalaropes with their tags, though, the scientists first had to catch them, which was no easy feat, as phalaropes often spent their time swimming far away from the shore or in shallow areas that were inaccessible by boat. They had the most luck with males guarding nests at Chaplin Lake.

When I first met Carle over Zoom in October 2023, he had recently seen the results from the first twenty-five tagged birds. They had surprised him. "Our understanding for thirty years was that they go to Great Salt Lake and Mono Lake and they eat a lot, and then they fly straight to Ecuador over the ocean," he told me. (In Ecuador, the salt company Ecuasal's man-made evaporation ponds host some of the world's highest densities of phalaropes.) But not a single one had traveled as expected. Instead, on the way to Ecuador, they had made stops at smaller intermediate sites: "coastal lagoons in Baja California and interior salt lakes in central Mexico and, like, little crappy ponds on military bases in the Mojave Desert," he explained animatedly. Those sites had never been on scientists' radar.

Once the birds arrived in South America for the austral summer, Castellino helped organize the first ever coordinated survey of phalaropes' non-breeding range, which was held in February 2020. It was a daunting prospect: the scientists had to survey everything from small salt flats in remote valleys high in the Andes to large lowland salt lakes like the Mar Chiquita, tailoring protocols to each site. The phalarope working group members joined forces with the Grupo de Conservación de Flamencos Altoandinos, or the High Andes Flamingo Conservation Group, whose members knew the terrain well from conducting annual flamingo counts for nearly three decades. (Salt lakes are also flamingo habitat; the birds get their pink color from eating brine shrimp.)

Once they had data from both the North and South American surveys, the working group estimated that the entire Wilson's phalarope population numbered one million—a one-third decline from the 1988 count of 1.5 million. Carle pointed out that it's hard to compare the figures because census methods have changed so much. But the more data they collected, the more a downward trend seemed clear.

Meanwhile, things were only getting worse for salt lakes. The following year, Abert dried up again, and Great Salt Lake reached a record low. When it did, the Center for Biological Diversity—an environmental organization well known for its use of litigation to protect imperiled species—began formulating a legal strategy to save the lake. They had three strong tools at their disposal, they realized. These were the same tools that Brigham Daniels later identified in his article: the Clean Air Act, which had obligated the LADWP to mitigate the dust spewed by the Owens lakebed; the public trust doctrine, which had mandated a water level for Mono and which might provide the impetus for rewriting water law in Nevada; and the Endangered Species Act.

The Endangered Species Act hadn't yet been used to conserve a salt lake, but its potential was enormous. A successful petition for a salt lake bird wouldn't protect just one species or one lake, but many of them, because it would come with federal enforcement to conserve habitat anywhere phalaropes stopped.

The organization considered filing a petition for either the Wilson's phalarope or the eared grebe. Those were the two species that relied on Great Salt Lake to such an extent that losing the lake would be "game over" for the species, in the words of Patrick Donnelly, the organization's Great Basin director. The decision became clear once Donnelly met Carle. "We had a young, hungry expert who was deep in the species and willing to put his name on the petition," he said.

Word of the petition began to spread in Utah in fall 2023. Some environmental organizations expressed their support without hesitation. Others, including the state governor, raised their hackles. They didn't want federal intrusion into Utahn matters—especially for an environmental cause. Part of this inhibition had to do with the influence of the Church of Jesus Christ of Latter-day Saints on Utah culture and politics. Though just 40 percent of Utahns now identify as LDS, the state governor and 90 percent of the state legislature do.

In the decades after World War II, the Mormon Church joined many other Christian denominations in shifting its politics rightward. In part, this was a reaction to changes in the culture of the American West. The region was shifting from an economy of ranching, logging, and mining to one of office jobs, tourism, and recreation. During the 1970s and 1980s, standoffs against federal officials as part of the Sagebrush Rebellion and debates over the Endangered Species Act—specifically, over the listing of the spotted owl—polarized the region into culture war.

The church, long a political force, took a side. The conflicts had reopened old wounds. In the 1800s, when Utah was a territory, the federal government had cracked down on polygamy, seizing property from the church, locking up officials, and denying statehood until the church agreed to renounce the practice.

"The resentment of federal power in the West and the idea that the federal government doesn't leave us alone—that's not distinctive, it's shared across the West," historian Matthew Bowman told me. "But for Mormons, it goes deeper than that, to the resentment and fear of the federal government going back to the 1800s. There's still a deep desire to be left alone, a sense of fear that the government is going to come in and tell us what to do—and it dovetails with general Western conservatism."

Those longstanding tensions made fights over federal environmental regulation particularly heated in Mormon country. In July 1999, for instance, a bishop in the town of Escalante, Utah, called for a "religious war" against environmentalists, and the town's Pioneer Day parade, commemorating the arrival of Mormon settlers to Utah, included a float where a dummy environmentalist was pinned to the grill of a car, as though run over. Similarly, it seemed, some didn't take kindly to the idea of the Endangered Species Act being deployed on their turf.

I spent the winter of the petition's announcement in Provo, the small city where church-affiliated Brigham Young University is located. During that time, I sat in on a senior capstone course about the Great Salt Lake taught by environmental science professor and lake activist Ben Abbott. During one course meeting, he quipped, "the most likely scenario is Utah figures out a way to get water to the lake. The second-most-likely scenario is federal intervention—they come in *E.T.*-style with helicopters and blow up the reservoirs!"

Abbott seemed torn between allegiance to the science and to

the state. As we drove his lab's electric car from the state capitol in Salt Lake City back to Provo one afternoon, I asked him about it. "There's an intangible reputational cost if it's not Utah in charge of the lake's rescue," he answered, thoughtfully choosing his words. "We love to pride ourselves on what we call the 'Utah Way' of doing government that isn't as acrimonious. This would represent a failure of the Utah Way—we couldn't get our act together, we couldn't save enough water."

Nevertheless, he signed the petition. Supporting it, he said, was a decision based on science. Not doing so would have been a decision based on politics.

A FEW MONTHS EARLIER, I'd finally gotten a pair of binoculars. I used them to spot birds from my desk in Tucson: Anna's hummingbirds sucking nectar from the potted lavender in front of my window; the vermilion flycatcher who guarded our block; mockingbirds, Gila woodpeckers, yellow-rumped warblers, and the rare cactus wren. Occasionally a Cooper's hawk brought prey to the tree in our front yard and eviscerated it in front of me.

One day, as I walked with Mariana in a nature preserve outside the city, she pointed to something perched atop a tall saguaro.

"Look at that one," she said with wonder. "It has colors!"

I flicked off the binoculars' rubber caps and pulled the lenses to my eyes. The blue and red pattern was unmistakable: it was a kestrel.

The more I paid attention to birds, the more lively the world felt. The birds themselves were light and full of color, endowed with the miracle of flight. But observing how they used the environment also gave me a kind of intuition about biodiversity. I took notice of the places where bird calls multiplied and more species revealed themselves. I observed their relationships with different plants, and the insects that served as pollinators and prey.

The birds brought the workings of the ecosystem to life for me in the way ranching once had.

I found myself thinking of French Christian philosopher Simone Weil. Weil wrote that "the authentic and pure values" of human life are truth, beauty and goodness. For her, religion was the route to reach these concepts, and attention a key spiritual practice in doing so. "Attention, taken to its highest degree, is the same thing as prayer. It presupposes faith and love," she writes.

That idea, I think, speaks even to those who do not think about God much or subscribe to any religious institution. Cultivating a practice of attention is a way to replicate the ambition and transcendence of religious thought, to share in the register in which concepts can be considered most fully.

Birds demand attention, and they reward it with a vitality so ethereal it feels like it must be something spiritual. I wasn't the only one to notice it. When I started going birding in high school, the most played track of my iTunes library was Sufjan Stevens's "The Lord God Bird." It's an unreleased song composed to go along with a National Public Radio story about the reappearance of the ivory-billed woodpecker, a black and white bird with a red pompadour and a white beak that for sixty years scientists had considered extinct due to logging. In 2005, there was a sighting of the bird, and NPR had commissioned Stevens to commemorate it. A friend sent the song to me when it came out.

I listened to it over and over. It starts off with the main motif, a mournful tune plucked on a banjo. When Stevens's soft voice comes in, conjuring the divine-seeming bird and the immense hardwood bayous of its final sighting, the combination of the minor key and the hopeful crooning feels transcendent, like a warmth you've never experienced before or knew could exist.

I'd first learned of Stevens on a music blog the summer before

                                                    SALT LAKES

high school. He'd already released a couple of albums by then, each with a mystifying and alluring mix of American regionalisms, cryptic confessional lyrics, and Christian belief. My favorite song was "The Predatory Wasp of the Palisades Is Out to Get Us!" Sung in the first person, it told a story about swimming and feeling a strange and unexpected desire for a friend.

Only years later could I attempt to articulate what I loved about the song. As the narrator suppresses his love for his friend, the desire takes on a life of its own, expanding to fill the whole landscape around them and turning to holy language to do so, with the chorus chanting motifs from hymns—"Hallelujah" and "Lamb of God"—at the same time as it praises rivers and landscapes and repeats, "We were in love." The solo voice becomes a full orchestra, as though the feeling is so certain of itself, so certain of its sacredness and beauty, that it swells to fill the space, even as the speaker can't act on the desire. I basked in the feelings it provoked inside me, a soaring sense born of its combination of nature, spirituality, and queerness.

At first, I didn't make the connection between the birds I saw in the mornings through borrowed binoculars and the song I hummed as I did my homework. As Stevens's career continued, however, birds appeared more and more in his songs. He explained in a 2017 interview: "I've always been obsessed with the sacred and with the profound, and I think birds are these divine creatures, because they can escalate to the heavens . . . birds feel like symbols of absolute freedom and transcendence."

That ascendant sacredness, birds' flashes of color and joy brightening the mundane, came to feel for me like an explanation for how queer love felt so surprisingly natural, and so surprisingly holy. Queerness and biodiversity: both forms of difference that enrich the world.

As the Utah legislative session came to a close in late February, I flew to Argentina to attend a meeting of the International Phalarope Working Group, joining thirty scientists from the length of the phalarope's migratory path—Canada, the United States, Ecuador, Bolivia, Chile, and Argentina—in Castellino's hometown on the shores of the Laguna Mar Chiquita. They discussed ways to standardize population counts across lakes, traded experiences and tips, and discussed future research and monitoring needs for the birds. The Great Salt Lake's two Johns flew over the lake with the Argentine park ranger who had conducted their aerial counts since 1992.

The town was beachy and touristic, featuring an esplanade lined with hotels, restaurants, and bars. A "flamingo train"—a flamingo-shaped open-air car that pulled two additional passenger cars—drove tourists on loops around the town and the surrounding lakefront. Unlike the Great Salt Lake, the Mar Chiquita has high-level protection: in 2022, nearly the entirety of the lake became a national park.

Such a designation was once on the table for Great Salt Lake. In 1967, Congress considered a proposal to turn the lake's Antelope Island into a national monument. But Utah senator Wallace F. Bennett wrote to President Lyndon Johnson that when compared to Utah's other national parks, it was "nowhere near their class in any way, shape or form" and referred to the site as a "barren desert island." Many Utahns, it seemed, agreed, and the plan failed. Some years later, a state senator suggested that the island be turned into a colony to which to exile people with AIDS. A freakish habitat for freakish people, but not in the fond way.

In its diversity of plants and birds, the ecological importance of the Mar Chiquita was clear. Yellow cuckoos—locally called *pirinchos*—and masked gnatcatchers, *taquaritas azules*, perched on subtropical jacaranda trees. Saffron finches, or *jilgueros*, landed on the long stems of pampas grass. There were flamingos! There were also familiar sights. The Utahns were excited to find pickleweed (*Salicornia spp.*), an edible plant that also grows along the shore of the Great Salt Lake. In Argentina it is known as *jume*, while Mariana knew it as *espárrago del mar*, or sea asparagus.

The scientists added to their bird "life lists" during the two-block walk from the hotel to the meeting's location at the town's municipal multipurpose room. Many spent their free time on the outskirts of town, looking for more birds. I tagged along with my binoculars, feeling like a kid—eager but unable to contribute. They had studied the species before arriving so that they could identify even birds they'd never seen in person before. When I joined a group one morning, we counted fifty-five species. It felt shocking to imagine that such biodiversity could be so close at hand.

We saw phalaropes only from a distance, stationed on the opposite shore in large groups, then passing us in the air like big clouds, glinting gray and white as they opened and closed their wings. The way they moved as a collective reminded me of the fluttering ascendance of the predatory wasp song, where the feeling of sacred warmth came from the unexpected match between queer desire and natural beauty. They're queer birds, the butches of the bird world, tough women who refuse to follow the rules. When they fly past you, though, they just look synchronized and delicate. It's that kind of unimaginable match that makes queer love so powerful. It defies common wisdom about what's correct, overwhelming you with something so beautiful that it turns the whole landscape sacred.

On a cloudy, windy day at the end of that month, Donnelly, Carle, Abbott, and Williams gathered on the steps of the Utah capitol for another rally, this time to launch the Endangered Species Act petition. They'd named the event Phalarope Phest. The podium held a large jar of Great Salt Lake water. As they had in January, the participants donned artist-made phalarope wings over their winter clothes and held long strips of canvas printed like waves with blue cyanotype. A trio of musicians on soprano saxophone, accordion, and tambourine led a bespoke rendition of "When the Saints Go Marching In" that referred to phalaropes' unusual feeding style: "Oh, how I want to be in that water / when the phalaropes come spinning in." The participants sang, danced, and cheered.

Donnelly emphasized the effectiveness of the Endangered Species Act. Scientists estimate that since 1973, the act has saved over two hundred species from going extinct. "I can't think of another government program that has a 99 percent success rate," he told the crowd. Abbott, invigorated by the international collaboration between Great Salt Lake and Mar Chiquita, told the crowd that the federal overreach story was incomplete. "The only way the state is giving away its sovereignty is if it denies there's a problem."

When Terry Tempest Williams took the podium, she called the listing "an act of love." It echoed her 2018 essay "A Totemic Act," in which she wrote that the Endangered Species Act was "both a policy and a prayer."

"The great consequence of the Endangered Species Act, over time, is that it ensures that we, as a species will not be alone," she writes in the essay. "We will remain part of a living, breathing, thriving community of vibrant beings with features, fins, and fur; roots, petals, and spines; trunk branches, and leaves." She told my co-reporter at *High Country News* magazine that "This is an ecological issue, an economic issue and a political one, but ultimately this is a spiritual issue."

There are traces of the Mormon spirituality in which Williams was raised in that vision of the connectedness of all life. In his late writings, church founder Joseph Smith became more expansive and experimental. He wrote that all of creation—animals, plants, and even rocks—shared a spirit that preceded its physical existence. He described elements of nature with human feelings and actions, like the earth groaning and speaking, and referred to both humans and nonhumans as "living souls." Brigham Young, who took over the church leadership from Smith, wrote in the same vein: "There is not one particle of element, which is not filled with life . . . There is life in all matter, throughout the vast extent of all the eternities; it is in the rock, the sand, the dust, in water, and gasses."

Over time, those teachings fell into obscurity. Thinkers like Williams and George Handley, the Brigham Young University English professor, are recovering them. They resonate with the beliefs about interconnectedness held by the Indigenous people of the West, too. "The land has a voice, and the lakes too. Nobody's listening to them, and they're going dry," Warm Springs Paiute Tribe elder Wilson Wewa told me.

The week I visited Abert, I was reading William Kittredge's *Hole in the Sky*, a memoir about growing up on his family's immense ranch in the desert of southeastern Oregon. As in *Owning it All*, he writes elegiacally about how rich and vital the landscape was during his childhood—before its wetlands were drained for crops and before trucks dominated ranching—and about the feeling of betrayal that he and his family experienced when they realized that the modern methods they'd been instructed to use were stripping the country of what they loved about it.

In the final pages of the book, he considers the possibility that his family's ranch would be accumulated into a large proposed protected area. It would mean erasing the years of labor he and his family had given to the country. But he was prepared for it. He knew it was the right outcome.

> *The redwood weirs my father built, the drainage canals and pumps, may be torn out, and the swamps may refill with spring runoff, to evolve back into tulebeds, and the waterbirds may come back in flocks beyond numbering . . . Our years of labor would have come to nothing, and that would be fine with me. I have a stake in the outcome. I want to ensure that my children's grandchildren can walk out into spring mornings which reek of life like the ones I knew.*

The passage reminded me of something scientist Dave Herbst told me about the importance of saline lakes: that saving them isn't about the dust, or the birds, or life on other planets, or any single thing. It's about abundant life. "We conservationists have to take the position of being biophiliacs—that we love life," he told me. "With less life, we're missing out."

I can imagine the mornings thronging with life that Kittredge describes because of what I've seen at salt lakes: gatherings of thousands of avocets, stretches of ground buzzing thick and black with alkali flies, mosaics of colorful tropical birds and delicate white shorebirds. But my sightings of life have always come with the knowledge of how much that presence of life has been diminished. The number of breeding birds on earth has decreased by three billion since 1970. Imagine those missing birds, and all the insects they would eat, surrounding us on earth.

The contemporary American West, and American hemisphere at large, is a place where colonization tried to violently eradicate forms of perceiving the world, living in it, and believing aspects of it to be sacred—and to replace them with others. It was a religious project and a military one. But Indigenous existence and belief weren't extinguished. Hundreds of years later, the settler colonialism of the West is unlikely to end anytime soon. Multiple cultures and belief systems, both religious and secular, inhabit the region. Astonishingly, there are certain things held sacred across them. Birds—"prayers with wings," as Williams called them—are one.

Birds draw us to them because their sacredness is so apparent. They bring color and music and strangeness to the world. They flit between the earthly and ethereal planes. Paying them attention helps us to see the way all the other beings that surround us are sacred, too—flies and shrimp and halobacteria and salt lakes.

Seen that way, the Endangered Species Act is more than an effective legal tool for saving salt lakes. It is an act of attention that protects the vitality that is the spiritual core of the world. We have to see the world in that register—the register of contemplation, of belief, of values bigger than politics. That was the register I was always seeking in my studies of literature. It's the only one in which the fellow beings of our surroundings matter enough for us to make the decisions that will save them for the future.

THE EPHEMERAL FOREVER

Lake Mackay, Western Australia/ Northern Territory, Australia

IN *SALT ON MINA MINA*, A 2007 PAINTING BY ABORIGINAL artist Dorothy Napangardi, white dots make off-kilter lines of varying widths and spacings that form, all together, an askew grid. From afar, the work looks like an old Sanborn map, revealing the alleyways, tenements, yards, and boulevards that make city grids imperfect, or a satellite image, like the pictures of Manhattan at night. Yet it's more animated than either of those. The lines pulse and flow. Their vibrations envelop the viewer.

I saw *Salt on Mina Mina* at the Musée des Confluences in Lyon, France, in the summer of 2019. I was on a trip to see two close friends, one of whom was finishing an apprenticeship to become a winemaker. Mina Mina, I later learned, is the name of the region of the Tanami Desert surrounding Lake Mackay, or, in the Walpiri language, Ngayurru. With a surface area of over 1,800 square miles, Mackay is the largest of hundreds of ephemeral salt lakes scattered throughout western Australia, and the country's fourth-largest lake. It fills only when cyclones cross the desert. Salts bloom on the surface of the ground as the rainwater evaporates, producing a reflective white surface that glitters.

"Most of the time it's dry. But when it's filled with water it's really beautiful, like a paradise," Walpiri activist Jeannie Herbert Punayi Nungarrayi said of it. "There are so many shells, water animals and waterbirds hovering around the lake. Like seagulls and other seabirds, beautiful black swans and kestrels and things like that."

Dorothy Napangardi was born in the early 1950s in Mina Mina, a landscape of spinifex, sandhills, and clay hardpans. Her family lived on lizards, bush apples, berries, and seeds. She had little contact with other people until she was around ten, when the family was forcibly moved to Yuendumu, a settlement founded in 1946 by the Native Affairs Branch of the Australian government. The family was unhappy in the town; they attempted to walk back to Mina Mina at night but were apprehended.

In the early 1980s, Napangardi moved to the town of Alice Springs with her first husband. There, she began taking classes at the Centre for Aboriginal Artists. Though she had received little formal schooling, she had been taught Jukurrpa—known in English by the imperfect early anthropologists' translation "Dreaming." Cultural studies scholar John Frow describes Jukurrpa as "a complex story of ritual, dance, body decoration, and song, and a temporal structure that is at once ancestral and present, transcendental and yet immanent in daily life." Taking place simultaneously in the present and the past, it recalls an epoch in which the land was inhabited by ancestral figures, many with supernatural abilities, who dictated the rules of life for Aboriginal people. In the words of Nungarrayi, Jukurrpa is "an all-embracing concept that provides rules for living, a moral code, as well as rules for interacting with the natural environment."

Certain Dreamings—stories pertinent to one's country and their associated dances, body decoration, symbols, and ceremonies— are "owned" by a group, and the stories passed on only within it. Some are strictly "men's business" and some "women's business."

That of Mina Mina belongs to the Walpiri women of the Napan-gardi and Napanangka skin groups. But because she had left Mina Mina when she was very young, Dorothy Napangardi didn't have the right to paint it: with only firsthand child's knowledge, the country and its dreaming did not yet belong to her.

In 1997, Napangardi took her family on a trip to visit Mina Mina so that she could be given knowledge about the site's Karnta-Kurlangu Jukurrpa, or "women-belonging Dreaming." The Dreaming describes the journey of a group of creator-women of all ages who travel east, starting from the place at Mina Mina where *karlangu*—or digging sticks, which are associated with women in Walpiri culture—emerged from the ground. The women take the sticks and dance their way east, using the digging sticks to create the landscape as it is now. Today, a stand of desert oaks marks the place where the ancestral karlangu emerged from the ground.

After the trip, Napangardi's painting changed. She stopped using traditional iconography and began painting only in black and white, creating the large canvases of dots-into-lines-into-wavering-grids that she would paint until her death in a car accident in June 2013. They are simultaneously narratives of ceremony, recreations of the process of creating the landscape, and landscape painting from a perspective on the world unimaginable to Western viewers. The white dots resemble the trail of the women's dance as well as the lines of crystallized salt. But then again, it was the ancestral women who created those salt lines in the first place.

ONE DAY, as I drove with Mariana in the Great Basin, she saw the bright reflection of a white disk of salt crust from the highway. "I want to go there," she said. So, days later, we packed a lunch and drove along the empty highway to what seemed to be the nearest point. We parked and set off, shuffling slowly through the sandy soil and sagebrush. It took longer than we expected to walk the three miles, because there was no path.

As we finally approached, Mariana saw an old water trough and decided it would be our lunch spot. We climbed in, leaning uncomfortably against the metal sides, and spread out our Tupperware. We had sandwiches, chips, fruit, and, as a treat, slices of cherry pie. She squeezed lime over the chips. The sun beat down on us.

After lunch, we explored. It was hard to tell where the lakebed began. First, the dry ground became an area of waterlogged ground lush with green saltgrass. Then, that marsh faded into soft crusts that looked like cauliflower florets. They collapsed under our feet, sending our boots deep into mud. That surface eventually gave way to a thicker, dryer crust, with a texture that looked like flaky pastry dough.

The lake had no clear shoreline because it was an ephemeral lake—one that fills only temporarily, at certain points in the year. Its playa had been drying, from the inside out, all summer.

The word *ephemera*, derived from an ancient Greek word meaning "lasting for only one day," appears all over the natural world. Salt lakes like Mackay, which fill only with large storms, are ephemeral lakes. In the American West, the 4.9 million miles of ephemeral streams, active just four to forty-six days of the year during periods of snowmelt or after rainstorms, are responsible for 80 percent of river flow. The mayfly family is named Ephemoptera, "ephemera with wings." Ephemeral plants are those with lifespans shorter than a growing season. In order to be able to complete their life cycle in under eight weeks, they are small, have shallow roots, and grow fast. In the southwestern US, desert ephemeral forage includes filaree (*Erodium cicutarium*), red brome (*Bromus madritensis*), and fiddleneck (*Amsinckia menziesii*). In Australia, button grass (*Dactyloctenium radulans*), whose inflorescence of three to ten spikes looks like the segments and vesicles of an orange, is known as "the most ephemeral of ephemeral grasses."

When I saw Napangardi's painting, the ephemeral was on my mind for another reason. The word is everywhere in queer theory, present in the specter of violence, the memory of AIDS, the fleetingness of hookups, and, most of all, dance. Critic Marcia Siegel wrote in 1972 that "dance exists as a perpetual vanishing point. At the moment of its creation it is gone." In 2009, queer theorist José Esteban Muñoz responded to Siegel, resisting the vanishing point. "Queer dance, after the live act, does not just expire," he wrote. "For queers, the gesture and its aftermath, the ephemeral trace, matter more than many traditional modes of evidencing lives and politics . . . after the gesture expires, its materiality has been transformed into ephemera that are utterly necessary." Dancer and writer Lydia Okrent agreed, writing in the introduction to *Mariana Valencia's Bouquet*: "Dance always leaves traces."

As I began to live queerly, I realized why dancing was so fundamental to queer theory. Dance introduced me to my body, giving me permission to occupy it fully. Dressed like the boys at beginner ballet class, I learned a new language of aesthetic strength; at La Cañita, surrounded by glitter and tattoos, I released it into space. But I still wanted to know: Why is the ephemeral such a crucial force in living a queer life? And how do you hold on to those traces?

WHEN I WAS WORKING as a ranch hand, I was invited to a retreat for women ranchers. We met four times, gathering annually for a long weekend at a ranch in California. The first year, everyone showed up baring tanned, sinewy arms, and qualified their introduction by saying that they hadn't wanted to come hang out with a bunch of women, they never hung out with women, that was why they worked in ranching in the first place—that it was only because of the insistence and generosity of the woman who had organized the retreat, a personal friend of nearly everyone there, that they had relented. I *had* wanted to come, but for the same reason: I wanted to find women who felt like I did about being a woman.

Slowly, everyone started to crack. They'd never had the space to talk with others about the feeling of living alone within male structures, the isolation of being with their husbands and their husbands alone, day after day. One woman talked about how her dog had lost a leg because her husband hadn't taken her seriously when she told him not to let it ride on the back of the truck on a muddy day. Another said she didn't feel like there was space for her to feel sad when she and her husband found dead animals. Soon, we all had to leave. It seemed to me a tragedy of heterosexual female friendship; by the time they had processed all the things that desperately needed talking about in their relationships with men, they had run out of time and had to go back to being wives.

The third year, one of the members gave her update: "I turned forty, my partner moved in, we got married, and she's pregnant." She had remained on the outskirts of the group even after everyone else had bought in, despite the fact that everyone respected, even adored, her. Her reserve made me curious. In part, it stemmed from the fact that she raised poultry and pigs instead of cattle. But it also seemed to indicate some important distance between womanhood and queerness. Then she added: "Getting married, having a family—I never thought it would be possible for me." Where many of the group's women struggled to be happy with their domestic lives, felt overwhelmed by the mountain of responsibilities that came from maintaining a family and a ranch, she radiated gratitude for and pleasure in what she had.

I took it as a kind of permission. For so long, haunted by the question of living sustainably in arid lands, ranching—its physical contact with the landscape, the way that dirt and hail and hide and blood layered on my skin, the way long hours turned slowly into an ability to see fecundity in what others saw as desert, a knowledge of the pieces of the landscape down to the individual species of grass and an intuition of how they work together, the opportunity to feel change, in my body, as both season and cycle—felt like an answer to my question of how to live rightly. The new approach of ecosystem thinking promised to be a way of making peace with destructive settler legacies. It also held a promise of roots, a connection to a piece of country so deep that a sense of home was undeniable. But ranching was a settlers' land use, part of the same system that has upended everything familiar and dependable, and much of what is pleasurable, about the world we live in. I needed a smaller life, in which significance wasn't measured in scale but depth. Like the Bidart line: "not only large things, a family, a book, a business; but the shape we give this afternoon, a conversation between friends, a meal."

When I was starting to let myself consider a queer future, I had copied the phrase in thin, all-caps handwriting onto a piece of cardstock and set it on the windowsill in front of my desk. It was how I wanted to live: a small life whose permanence was generated out of attention to quotidian ephemera. A way of being that, to me, was instantly recognizable as queer. A queer ecology, in the sense of the Greek root of the word: a queer household.

For me, each banal act of daily life is part of the specificity of queer life. Spending Sunday morning sweeping the house. Biking to the market and coming back with backpacks full of produce, imagining what we can make in the coming week. Sitting at the kitchen table de-stemming the purslane. Trading off making lunch. Hanging the clothes out to dry. Mariana dragging logs around in the backyard wearing boy's swim trunks and a cashmere sweater.

Accompanying each other on errands just to spend time together. Jumping in the shower after swimming. Giggling about memes.

Some of the most beautiful, sacred parts of love are the banal ones, just like some of the most beautiful and sacred parts of the natural world. In the rush of heterosexual life, those things were chores. They felt like time unfairly taken from me by gender roles. But in being queer, the quotidian became the substance of real life—not a tedious backdrop to what was meaningful, but the whole point. A quotidian act, Muñoz writes, "signifies a vast lifeworld of queer relationality . . . and a utopian potentiality." The joy of queerness is in the richness of the quotidian, just as the pleasure of the world is in vital and abundant surroundings.

In the water trough, Mariana and I ate grocery store food out of containers we'd used hundreds of times and would use hundreds more, wearing the same clothes and talking about the same things as we would on any ordinary day. But the world around us became suddenly extraordinary, like something that could only exist in mythology or metaphor. I understood why the Mexica people had named the Mexico Basin the navel of the universe: in the playa lake's small, open valley, I could see how the water from all the surrounding canyons flowed to the tiny round recess where we were sitting, and how the little notch was dramatically embraced by the mountains, the sky, and all of space. As in Dreaming, where the world is neither that of the present nor that of the ancestral past but simultaneously that of both, ordinary time was suspended. The ephemeral and the eternal interlocked.

SCIENTISTS PREDICT THAT climate change will make ephemeral lakes even more ephemeral: they'll stay dry for longer periods, will fill less frequently, and will desiccate more quickly when they do, leading to a decrease in biodiversity. Elsewhere, as at Abert, diversions and drought are turning perennial lakes into ephemeral ones. Climate change is also making things more unpredictable. In spring 2024, a series of hurricanes and atmospheric rivers—corridors in the atmosphere that carry moisture from the equators toward the poles, swollen with overheated vapor—dropped long, hard rainstorms on California. They filled Badwater Basin, the main basin of Death Valley, three feet deep with water, ephemerally reviving its lake, Lake Manly.

Several of my friends drove to kayak on it. One, the writer Miles Griffis, wrote afterward that as he floated on its surface, he recalled the artist David Wojnarowicz's visit to Badwater Basin in 1991, a year before his death from AIDS. At the dormant lake, Wojnarowicz had photographed himself wrapped in a white sheet, staging a funeral and reclaiming a desert place much like the scorned, dry Antelope Island that others saw as fit for queers like him. For Miles, disabled by Long Covid and grieving the loss of health and of quotidian life as he had known it, Wojnarowicz's work had a searing resonance. "I let the water surround my face and mouth until I was just a floating oval of skin," he wrote. "I knew that in the coming weeks, the lake would shrink to a wet, muddy surface before disappearing as quickly as it had arrived. The lake wasn't just a mirror of the sky; it reflected the ephemerality of life."

That winter, Mariana and I drove to Mexicali, across the US–Mexico border from the Imperial Valley, and camped at the Laguna Salada, a kind of mirror image of the Salton Sea—another "sink," or basin, into which the Colorado River once used to spill as it coursed toward the Gulf of California. Friends told us it was possible to drive across the lakebed, but as soon as we set out on the two-track that bisected the playa, dust filled the car and we turned around, coughing. We took the route that circumvented it, crossing sandy fields of palo verde, California croton, and creosote—called *gobernadora*, or lady governor, in Spanish, for the way it prevents other plants from growing nearby—and camped in the mountains to its southwest.

The following day, we drove into the city to meet up with Jessica Sevilla, Mayte Miranda, and Rosela del Bosque, cofounders of the Archivo Familiar del Río Colorado. The *archivo* is a collective of artists and curators whose name is a play on the double meaning of the word—in English it translates to either the Colorado River Family Album or the Colorado River Family Archive. We drove south from the gray, low-slung city through the desert to the Hardy River, a tributary of the Colorado that drains agricultural runoff. They pointed out the riparian plants of the delta, including arrowweed, whose name in Spanish, *cachanilla*, is also a nickname for people from Mexicali. We climbed a hill and looked out over the landscape: the river, the communal fields, the industrial plants and geothermal power station they called the "cloud factory" for its tall white plumes. The mountain range to the west blocked our view of the laguna: the Indigenous Cucapah people's name for it, they told us, was "water behind the mountains."

The past inhabitants of the Mexicali Valley likely knew the Laguna Salada in a similar form as the one that Jeannie Herbert Punayi Nungarrayi describes Lake Mackay: mostly dry but glistening and filled with birds when the river chanced to fill it. Now the lakebed rarely, if ever, fills. That's why the Archivo Familiar exists. Its members collect photographs, anecdotes, and other quotidian evidence of the lost waterways of the region.

"It's beautiful to hear a story about how two blocks away, where now there's a highway, there used to be shrimp and a place you could fish," Sevilla told me. "Or that the Alamo Canal, which now is pretty much always dry, used to freeze over sometimes."

They turn those documents into artwork, exhibitions, and performances that make visible the landscape's cartography of loss—and imagine possible futures for those waterways. One, *Nuestra Señora del Río Oculto*, imagined an apparition of the Virgin Mary at the Laguna Salada with an accompanying performance using sounds recorded at the dry lake. Others play with past, present, and future, overlaying photographs of former waterways with images of their present state, sometimes intervening in them with speculations and questions.

After Mexicali, Mariana and I drove from the borderlands to Mexico City, passing the expansive ephemeral rangelands of Sonora, watching tropical forest gradually engulf the cacti as the road took us south through Sinaloa, cresting the mountains of Nayarit and crossing the industrialized highland plain of Guanajuato and Querétaro. It was a spring of record heat and drought in the country, and three of central Mexico's endorheic lakes had dried up. In Michoacán, brigades of locals went to clean the springs that surrounded the Lago de Pátzcuaro. The nearby Lago de Cuitzeo, which we had driven along the year before, its blue and green landscape rich with cattle egrets and black-necked stilts, was an expanse of light-brown dirt. And when the government had canceled the construction of the New Mexico City International Airport on the Texcoco lakebed, they'd instead built on the bed of the Lago de Zumpango, at the north end of the Mexico Basin. What remained of the lake went dry in the process, along with springs all across the aquifer.

When we arrived, the city was tense with water crisis. Trucks trundled through the trafficked streets bringing expensive water even to neighborhoods that had always been favored by the city's grid. Ours received water just a few hours a day, and the buzz of the building's aging pump turning on became a new imprint on our daily lives, sending us to fill as many buckets as we could. In the hot, dry May, searing in a place long considered to have perfect, springlike weather all year, a place where Mariana remembered having to wear a scarf and mittens on winter mornings as a schoolchild but where you could now bicycle in a T-shirt in December, where the rainy season had begun coming later in the spring but with greater force, bringing hours-long thunderstorms with droplets everyone considered to be enormous, I thought about how everything felt so different, different from just a few years ago, different from when I was in high school, different from when I was a child. In our lifetimes, climate change has made the whole world feel ephemeral.

WE ARE LIVING IN a time of loss. Even if we manage to renovate the working country of the West by buying water rights from willing sellers, even if litigation over the public trust doctrine transforms the minutiae of our recalcitrant water law, even if we restore lakes to the tribes that have long stewarded them, even if we list the Wilson's phalarope and eared grebe as threatened and enforce a strict agenda to conserve the habitat they depend on, even if we stop seeing salt lakes as sterile and useless and treat them instead as sacred and vibrant, so much destruction has been in motion for so long that we will likely still end up living in a changed, muted world. Lakes and rivers will dry up; insects and birds will die off. They already are. Some scientists predict that global warming will cause the extinction of 70 percent of all species by this century's end.

Just as queer ecology teaches us to celebrate the ugly and the unruly of the world, the scorned and the sacred, it also prepares us for these losses. The presence of the ephemeral in queer life is many-hued. It trains us to grasp the flashes of utopian pleasure on the horizon, but also to mourn the kin unjustly taken away. "Queer and trans peoples' experiences of grieving premature

death," writes the hydrologist Cleo Woelfe Hazard in *Underflows,*
"can stoke collective action on extinction and repair." We have
to mourn to recognize the gravity of the losses that surround us,
and to feel how important it is to keep fighting to staunch them.

Napangardi said of her art, "I want my grandchildren to know
the country." Like the physical world, the time-space of Dreaming is a country full of ephemeralities—songs that dissipate into
air, paint washed off, dance that vanishes. Practicing ceremony
and ritual, as she did in learning Karnta-Kurlangu Jukurrpa as an
adult, is one way of grasping their traces. Dreaming has survived
violent transformations; it is said to have 50,000 years of history.
The permanence of art-making is another, a way of fixing the feeling of a memory in time.

When loss is unavoidable, there is solace in grasping the ephemeral. We grieve the fleeting and celebrate what remains. We catch
the dense instants of life and hold tight to their richness.

ACKNOWLEDGMENTS

Versions of these chapters appeared in *[PANK]*, *Drunken Boat/ ANMLY*, *New South*, *Kenyon Review Online*, *High Country News*, the *Guardian*, and *Shenandoah*. I'm grateful for those editors' support of my work, and for the important education as a writer I received from those with whom I worked early in my career.

During the decade I spent working on this book, I benefited from a variety of institutions, among them Headlands Center for the Arts, the Fulbright Commission, the University of California–Berkeley Department of Geography, Deep Springs College, PLAYA/ Summer Lake, Oregon State University's Spring Creek Project, and Brigham Young University's Charles Redd Center for the American West. Many thanks as well to Nick Baefsky and Amy Wright.

In 2022, Raquel Gutiérrez selected the essay that became this book's coda as the winner of the Waterston Desert Writing Prize. It was a vote of confidence I needed, and one that was all the more meaningful because it came from Raquel, whose work I had long admired.

The following spring, Bridget Matzie contacted me, partially because we shared an interest in the Salton Sea. I'm grateful for her immediate and unwavering enthusiasm for this project; she championed it in a way that previously I had only been able to imagine.

Helen Thomaides acquired the book for Norton, and her belief in it also earned my infinite gratitude. When Matt Weiland took over as this book's editor, he proved an ideal reader for it, and I feel lucky to have benefited from his kind and incisive guidance throughout the revision process.

Eve McGlynn patiently worked with me to tailor the maps that appear in the front of the book. Jennifer Gersten formatted the bibliography on a tight timeline. Finally, this book wouldn't have been possible without the time and knowledge shared with me by my many interlocutors. It was invigorating to learn from each of you.

The world in which I grew up—my family's habits and values shaped over generations—gave me both a sense of place and the curiosity to interrogate it. My love of reading, art, music, languages, and the natural world were seeded earlier than I can remember, and cultivated through opportunities to learn and to experience the world that I have only been able to fully appreciate as an adult.

One of the most beautiful aspects of queer adulthood is the way that life is shaped by friendship. During its long gestation period, this book was enriched by the companionship of Jill Bodner, Isabel Newlin, Aspen Reese, Shizue Roche Adachi, Sahiba Sindhu, Nazik Tynystanova, Katie Peterson, Young and Emily Suh, Jesús Gutiérrez, Chris Lesser, Julia Sizek, Nicolás Medina Mora, Whitney DeVos, César Cañedo, Jocelyn Saidenberg, Miles Griffis, and many others. Thanks to the friends who helped make Tucson into home: the members of SNAG, Robert Villa, Rachel Glickhouse, Eliseu Cavalcante, and others. Andrew Emery Brown took my author photos over the course of an evening walk downtown, in interruptions to a conversation about novels and water infrastructure.

Dylan Kenny accompanied me through my twenties, during which time we taught each other how to show and experience love. I'm proud of the years we spent together and moved by the friendship we have now. The Kenny–Perkins–Tressler clan was a second family to me for many years, and I'm grateful to still have a place in it as Alva's godmother.

When I first met Mariana, she told me that between the glyphs of the Mexico City metro and her collection of stickers in WhatsApp, society had moved past the need for writing. Así que en lugar de palabras te mando un sticker en forma de corazón ♥.

NOTE ON SOURCES

Prologue

The book that made me feel like a path had cleared to becoming a writer was Wallace Stegner's *The American West as Living Space*. I followed it with Philip Fradkin's biography of Stegner, *Wallace Stegner and the American West*.

In large part, I have my maternal grandfather to thank for my interest in the environment. In attempting to trace the origins of my interest in the connections between Russian literature and that of the Western US, I have numerous high school teachers to thank for shepherding my early interests, and later the professors who helped me deepen my skills as a scholarly reader without losing the pleasure and freedom of more poetic approaches.

The novels and stories I mention include Chekhov's *The Steppe and Other Stories*, translated by Ronald Wilks; Saltykov-Shchedrin's *The Golovlyov Family*, also translated by Wilks; Platonov's *Soul*, translated by Robert and Elizabeth Chandler with Katia Grigoruk, Angela Livingstone, Olga Meerson, and Eric Naiman; Tolstoy's *War and Peace* and *Anna Karenina*, both translated by Richard Pevear and Larissa Volokhonsky; and Pasternak's *Doctor Zhivago*, translated by Max Hayward and Manya Harari. Orlando Figes's *Natasha's Dance* and Christopher Ely's *This Meager Nature* were both instrumental in helping me understand Russian and Soviet conceptions of landscape and nature. While I was working on this book, Oksana Vasyakina's *Wound*, a queer Russian road novel whose thinking about space I relished, was released in Russian and later in English in Elina Alter's translation.

For the Great Basin, John McPhee's *Basin and Range* is a classic. While in residence at PLAYA/Summer Lake, I found Samuel Houghton's *A Trace of Desert Waters: The Great Basin Story* and joked to friends that it was the "straight version" of this book. Ed Dorn's *The Shoshoneans*, a gift to me from Dylan, is an idiosyncratic travelogue worth reading if nothing else than for getting to know the odd figure of Dorn, a little-known Western writer who was a student of Charles Olson and an early champion of Lucia Berlin.

As I tried to grasp the basics of salt lakes, Robert Jellison and William D. Williams's article "Salt Lakes: Values, Threats and Future" and Williams's "Environmental Threats to Salt Lakes and the Likely Status of Inland Saline Ecosystems in 2025" were my best guides. Conversations with Ron Larson, Patrick Donnelly, and Dave Herbst helped me untangle the differences between perennial, seasonal, and playa lakes. Wayne Wurtsbaugh's article "Perspective: Decline of the World's Saline Lakes," is an introduction to both salt lakes and their decline. Other useful articles for understanding the decline include Johnnie Moore's "Recent Desiccation of Western Great Basin Saline Lakes," Joe Eilers and Ron Larson's "Introducing Terminal Lakes" in *Lakeline*, and Michael Wine's research, including the article "Irrigation Water Use Driving Desiccation of Earth's Endorheic Lakes and Seas." The mention of drying lakes' contribution to sea level rise comes from "Recent Global Decline in Endorheic Basin Water Shortages" by Jida Wang et al.

Chapter 1: Closed Basins and Sacred Lakes

Lauren Steele's 2021 article "The Water That Couldn't Save," published in *Deseret* magazine, provides a thorough overview of the crisis of Great Salt Lake. Bonnie Baxter's "Great Salt Lake Microbiology: A Historical Perspective" surveys the history of scientific writing about the lake. The Great Salt Lake Strike Team's February 2023 "Great Salt Lake Policy Assessment" unites the perspectives on the lake's future of multiple key Utah scientists.

A video of the January 2024 rally to save Great Salt Lake is available on YouTube. To learn more about Darren Parry and Shoshone displacement in Utah, read his book *The Bear River Massacre: A Shoshone History*. Terry Tempest Williams's *Refuge: An Unnatural History of Family and Place* is timeless even as the climatic and political circumstances at the lake have changed. (It also inspired this book's subtitle.) Her essay "I am Haunted by What I Have Seen at the Great Salt Lake" is an important update. Ben Abbott's fight against islands in Utah Lake was covered by Leia Larsen in the article "A Utah Lake dredging critic got dragged into a lawsuit with a private company. Did he also lose a public grant?" The text of his speech at the rally can be found on his blog.

My article "LDS Environmentalists Want Their Institution to Address the Great Salt Lake's Collapse" appeared in *High Country News*. For it, I interviewed members of the Mormon Environmental Stewardship Alliance (MESA) and LDS Earth Stewardship, along with George Handley, Jason Brown, and John Larsen. Its textual sources included Jason M. Brown's "Whither Mormon Environmental Theology?," Richard Foltz's "Mormon Values and the Environment," George Handley's "Mormonism: Mormon Views of Environmental Stewardship," and Handley's essay "The Restoration of All Things: The Case of the Provo River Delta." (The quotation at the end of the chapter comes from this last essay.) Handley, Terry B. Ball, and Steven L. Peck's edited volume *Stewardship and the Creation: Perspectives on the Environment* is also a valuable resource. My articles about Utah Youth Environmental Solutions's Die-in for Great Salt Lake and my interview with members of the sixth-grade class that successfully lobbied for the brine shrimp to become Utah's state crustacean were also published in *High Country News*.

My introduction to the history of the Church of Jesus Christ of Latter-day Saints came from interviewing Matthew Bowman, author of *The Mormon People: The Making of an American Faith*. Benjamin Park's *American Zion: A New History of Mormonism* came out from Liveright, an imprint of Norton, as I was conducting research for *Salt Lakes*, and Helen Thomaides, this book's first editor, mailed me a copy. Leonard Arrington's book *Great Basin Kingdom: An Economic History of the Latter-day Saints, 1830–1900* is the key source for the economic—and therefore environmental—history of Mormon settlement in Utah and the West. The Wallace Stegner quotation about Salt Lake City comes from the essay "At Home in the Fields of the Lord," collected in *Marking the Sparrow's Fall*.

The Church of Jesus Christ of Latter-day Saints official style guide currently discourages the use of the term "Mormon" as shorthand for its full name. The guide offers, instead, the "Church of Jesus Christ" or the "restored Church of Jesus Christ." In his book, Benjamin Park writes that though the term "Mormon" was "originally born of derision," for most of the church's history, it evoked a sense of "cohesive identity." He uses the term as shorthand because, he writes, it was the term most frequently used during the history his book overs. This is also true for much of the history covered in this book. Additionally, as Park charts in his book, the renunciation of the Mormon moniker came in 2018, at a moment in which the church had decided to downplay its differences from mainstream Christianity, and emphasize the worship of Jesus Christ, in order to better ally itself with the Christian right. My book is, in part, about the beauty of beings, lives, and beliefs that fall outside the norm. I have respect for the uniqueness of Mormon scripture and culture and my use of the term "Mormon" carries this respect with it.

My understanding of the early Mormon settlers' relationship to the lake, the history of the relationship between settlers and Native peoples in the nineteenth century, and the twentieth-century management of the lake, came largely from archival research conducted at Brigham Young University and the University of Utah with the support of BYU's Charles Redd Center for the American West. At BYU's Harold B. Lee Library Archives and Manuscript division, the James Talmage Papers, Virgil Peterson Geological Papers, Willard and Celia Luce Collection, John Steele Collection, Wallace F. Bennett Papers, and Thomas Day Jr. Collection were particularly beneficial. At the University of Utah Marriott Library, I spent particular time with the Myron L. and Lucille P. Sutton Papers, the Thomas Caldwell Adams Papers, and the Brigham D. Madsen Papers.

My understanding of how Mormon settlement shaped Indigenous history in the Great Basin was greatly augmented by three sources: Ned Blackhawk's *Violence over the Land: Indians and Empire in the Early American West*, Jared Farmer's *On Zion's Mount: Mormons, Indians, and the American Landscape*, and Howard A. Christy's article "Open Hand and Mailed Fist: Mormon-Indian Relations in Utah 1847–52." Reading settlers' journals and correspondence between Brigham Young and US government officials in the archives helped me see the divide between the perception of church policy and the reality.

Marc Reisner wrote in *Cadillac Desert* that "no one was better at irrigation farming than the Mormons," that they "attacked the desert full-bore, flooded it, subverted its dreadful indifference—moralized it" and that their influence as irrigators was so great that the Bureau of Reclamation was "based on Mormon experience, guided by Mormon laws, run largely by Mormons." There are no citations for these claims, and I've come to find them rather fabulist. When I began interviewing church members about the Great Salt Lake, I found that most were familiar with Reisner's argument, but none had evidence for it either. So, part of my work in this chapter was to figure out just how Mormon irrigation influenced American society at large. In this effort, I drew on Ralph Hess's article "The Beginnings of Irrigation in the United States," Alexander Thomas's "Irrigating the Mormon Heartland: The Operation of the Irrigation

Companies in Wasatch Oasis Communities, 1847–1880," and Arrington and Dean May's "'A Different Mode of Life': Irrigation and Society in Nineteenth-Century Utah." *The Colorado Doctrine: Water Rights, Corporations, and Distributive Justice on the American Frontier*, by David Schorr, helped fill in the prior appropriation side of the story. Finally, conversations with Ben Abbott and Brig Daniels helped me shape my thinking about the role of the lake and the role of irrigation in Utah, Mormon, and Western history.

Chapter 2: Richness in Damaged Places

A very early form of this chapter appeared in *[Pank]* magazine in 2015, a year after I graduated from college. It was my first published essay, and a vote of confidence that propelled me to keep going as a writer.

The books I read that made me want to see the Salton Sea were Carey McWilliams's *California: The Great Exception* and *Southern California: An Island on the Land* and William Vollmann's *Imperial*. Later, Frank Waters's *The Colorado* joined the list. Mike Davis's *City of Quartz* also shaped my early months in California. My Central Valley reading list began with Joan Didion's *Slouching Towards Bethlehem* and grew during the years that I was with Dylan, including McWilliams's *Factories in the Field*, Richard Rodríguez's *The Hunger of Memory* and *Days of Obligation*, Heyday's *Highway 99* anthology, and numerous newspaper articles about pesticides, groundwater, high-speed rail, sprawl, and other topics.

Traci Voyles's book *Settler Sea* came out while I was writing this book and helped me add historical background about Native and settler relationships to the sea, and about the history of the wildlife refuge. William McLaren's article "A Fishery, a Sanctuary, a Sink, and a Disaster: The Often-Hapless Management of California's Salton Sea" helped me understand the Salton Sea Cases and the circumstances of the creation of its wildlife refuges. William duBuys's *Salt Dreams* is one of the most thorough sources on the history of the sea. I had generative conversations with Michael Cohen and Alida Cantor that helped me understand the contemporary regulatory geography of the sea. Julia Sizek, a friend and great font of southeastern California knowledge, came to visit me in Tucson and found a copy of *Mukat's People: The Cahuilla Indians of Southern California*, by Lowell John Bean, in a thrift store a short drive from my house.

The *New York Times* article in which I first learned of the sea was about the director Elgin James, who entered the Sundance Labs after serving a prison sen-

　　　　　　　　　　　　　　　　　　　　　　NOTE ON SOURCES

tence and went on to make the film *Little Birds*, about two adolescent girls from the shores of the Salton Sea who get an ill-fated ride to Los Angeles. I never saw the film, and remembered only faint details from the review, but I stored away the sea in my memory.

The article about the Indio date girls is KQED's "How the Eastern Coachella Valley Used Egyptomania to Sell Dates." Geographer Natalie Koch's *Arid Empire: The Entangled Fates of Arabia and Arizona* gives a fascinating and deep overview of Middle Eastern and Holy Land symbolism in the colonization of the southwestern US. KQED's "Plagues and Pleasures at the Salton Sea," online on YouTube, provides a visual foray into the sea's tourist heyday.

When I visited the sea, I'd heard about fish die-offs from the late Diana Marcum's reporting for the *Los Angeles Times*. More recent reporting by Ian James, Erin Rode, and Janet Wilson has done an important job of keeping tabs on the complex developments at the Salton Sea and the Colorado River. Wilson and Nat Lash's joint investigation, "The 20 Farming Families Who Use More Water from the Colorado River Than Some Western States," provides insight into the power of the farmers of the Imperial Irrigation District.

The Quantification Settlement Agreement, now over twenty years in the past, put the future of the Salton Sea into limbo. David Owen's *Where the Water Goes* explains the nature and the stakes of the water transfer agreement. The story continues to develop, with the federal government, Imperial Irrigation District, tribes, and environmental organizations all involved in bureaucracy and litigation to determine the sea's future.

As the sea has dried up, its dust has become a major public health concern. In 2023, I wrote about the Salton Sea Environmental Timeseries, a unique partnership, led by Ryan Sinclair of Loma Linda University, between academic and community scientists who are monitoring the sea's water and air quality in the absence of state or federal monitoring, in the article "In Search of Answers at the Salton Sea" for *High Country News*. Scientists including David Lo, Will Porter, and Trevor Biddle are all conducting groundbreaking work on asthma in the communities surrounding the sea that suggests that the inflammation caused by the dust released as it dries up is unique relative to conventional asthma pathways.

The sea continues to change in other ways, too. On one hand, it is attracting more and more tourism: the Bombay Beach Biennale, created by filmmakers, hoteliers, and influencers from Los Angeles, has made the declining resort town a destination again. On another, mining companies hope to transform the Imperial Valley into "Lithium Valley." Some community members have high hopes for the economic development this could bring. As with the sea, however, it remains to be seen how health, environment, economy, and equity will be balanced.

Chapter 3: Imperfect Recovery

I wrote the first draft of this chapter in the winter of 2015–16, while working as a cowboy. Given the isolation in which I found myself, its expository sections relied largely on information about the Aral Sea I gathered from news articles on the internet. These included the BBC's feature "Waiting for the Sea," *New Internationalist* "A Sea Returns to Life, a Sea Slowly Dies," and the *Encyclopedia Britannica* entries for the Aral Sea, Syr Darya, and Amu Darya. The version published here is updated with information confirmed using more recent sources.

I read Chingiz Aitmatov's *The Day Lasts More Than a Hundred Years* first in English, translated by F. J. French, and later in Russian in a set of the complete works of Aitmatov that I purchased in Bishkek. Another book on my mind at the time was Donald Worster's *Rivers of Empire*.

More information about the history of collectivization, including Uzbek resistance to it, can be found in Marianne Kamp's article "Soviet Collectivization in Central Asia" and her book *Collectivization Generation: Oral Histories of a Social Revolution in Uzbekistan*. On the cotton industry, David Tarr and Eskender Trushin's white paper for the World Bank, "Did the Desire for Cotton Self-Sufficiency Lead to the Aral Sea Disaster?" was a useful source of dates and statistics. NASA Earth Observatory also has numerous posts chronicling the decline of the Aral.

On the saiga die-off, the *New York Times* coverage is thorough: see "Death on the Steppes: Mystery Disease Kills Saigas," "More Than Half of Entire Species of Saigas Gone in Mysterious Die-off, and "A Warm and Wet Spring, Then 200,000 Dead Saigas."

Information about the health effects of the Aral Sea's decline can be found in numerous scientific articles, among them Philip Whish-Wilson's "The Aral Sea Environmental Health Crisis," Anjim Shamshetova's "Ecological Problems of the Aral Sea Region and the Psychology of the Region," and "Immunological Characteristics of Diseases of the Bronchopulmonary System in Infant Children in the Aral Sea Region" by N. R. Alieva et al. Information about Vozrozhdeniya Island can be found in the BBC article "The Deadly Germ Warfare Island Abandoned by the Soviets" by Zoria Gorvett.

On the topic of the Kok-Aral dam, the BBC also published "The Country That Brought a Sea Back to Life," by Dene-Hern Chen; *New Internationalist* published "Death and Re-Birth of a Lake: How Water Came Back to the Dry Aral Sea."

Additional sources about the Aral Sea from the field of environmental history include Christian Teichmann's article "Canals, Cotton, and the Limits of De-Colonization in Soviet Uzbekistan, 1924–41"; Julia Obertreis's *Imperial Desert Dreams: Cotton Growing and Irrigation in Central Asia, 1860–1991*; William Wheeler's *Environment and Post-Soviet Transformation in Kazakhstan's Aral Sea Region*, and Maya Peterson's *Pipe Dreams: Water and Empire in Central Asia's Aral Sea Basin*.

Chapter 4: The Common Good

During my research, Arya Degenhardt of the Mono Lake Committee graciously mailed me a copy of *Storm Over Mono* by John Hart. The book—now out of print, though available to read for free online—tells the story of the Mono Lake fight and was a key source for this chapter.

Tim Lowenstein helped me understand the geology and chemistry of the lake. For the Native history of the lake, my sources included Kat Anderson's *Tending the Wild* and Julian Steward's *Basin-Plateau Aboriginal Sociopolitical Groups*. Though Steward offers precise observations about Northern Paiute life, his "salvage anthropology" must be read critically. The William Brewer quote comes from *Up and Down California*, quoted in *Storm over Mono*. The quotations about settlers not being welcome to "sit down on the grass patches" and the tribes subsisting "entirely upon the grass seeds and nuts" come from Colonel George Evans, quoted in Bruce Pavlik's *The California Deserts*.

The Kootzaduka'a tribe have been seeking federal recognition for many years. More information about the tribe can be found on their website.

Robert Marks's articles "Dispossessed Again: Paiute Land Allotments in the Mono Basin 1907–1929" and "Before Mulholland: Land, Water, and Power in the Mono Basin, 1872–1923" were informative for understanding how settlement dispossessed Native Californians before the arrival of Los Angeles. Reading the Indian Claims Commission decision for the Northern Paiute Nation was instrumental in understanding how Paiute lands were transferred to the public domain, and eventually to private landowners or the Bureau of Land Management, after the Treaty of Guadalupe Hidalgo. The ICC decisions are available digitally on the Oklahoma State University Library website.

Joseph Sax's article "The Public Trust Doctrine in Natural Resource Law" originally appeared in 1970 in the *Michigan Law Review*. Alida Cantor's article "The Public Trust Doctrine and Critical Legal Geographies of Water in California" helped me understand the doctrine's history. The 1977 Mono Lake Study is available in the archives of the Mono Lake Committee at the special collections of the University of California–Riverside and online at the Mono Basin Clearinghouse; the Mono Lake Committee's archives are also held at UC–Riverside. I also drew on Joseph Jehl's article "Biology of the Eared Grebe and Wilson's Phalarope in the Nonbreeding Season: A Study of Adaptations to Saline Lakes."

I had the fortune to interview David "Bug" Herbst, David "Wink" Winkler, and Gayle Dana, who formed part of the 1976 summer research team behind the seminal ecological study—speaking with them was like a celebrity encounter. Bonnie Baxter explained the brine fly life cycle to me, and I spent a dreamy morning kayaking at Great Salt Lake with her and Dave Herbst. Finally, I spoke with Geoff McQuilkin, director of the Mono Lake Committee. The committee continues to do important work and we are fortunate that Mono Lake has such an institution to steward it.

The "shimmering bubble shield" reference comes from the PBS video "This Daring Fly Swims in Shimmering Bubble Shield." I learned about brine shrimp reproduction in Wayne Wurtsbaugh and Z. Maciej Gliwicz's "Limnological Control of Brine Shrimp Population Dynamics and Cyst Production in the Great Salt Lake, Utah." Shrimp and phalaropes each have brief mentions in Bruce Bagemihl's *Biological Exuberance: Animal Homosexuality and Natural Diversity*.

Chapter 5: Restitution

The Last Ranch: A Colorado Community and the Coming Desert, by Sam Bingham, focuses on holistic management, a grazing and general life-planning method based on the book *Holistic Resource Management* by Allan Savory. Holistic Management International, a company that provides trainings to ranchers based on Savory's methods, has grown in recent years at the same time as it has attracted controversy. The ranch where I worked adopted some of its principles, such as rotating the cattle through relatively small pastures as well as a general outlook of interest and concern for the ecology of the grassland ecosystem, but did not follow it as a strict program.

Further useful perspectives on the ecology of ranching include Diana K. Davis's *The Arid Lands: History, Power, Knowledge*, Nathan F. Sayre's *The New Ranch Handbook: A Guide for Restoring Western Rangelands* and *The Politics of Scale: A History of Rangeland Science*, and Hannah Gosnell's article "A Half Century of Holistic Management." On the history of ranching, Paul F. Starrs's *Let the Cowboy Ride* covers the West region by region.

Raymond Williams's line "a working country is hardly ever a landscape" comes from his book *The Country and the City*. I encountered the ideas of "significant otherness" and "becoming with" in Donna Haraway's *The Companion Species Manifesto: Dogs, People and Significant Otherness*. My New Mexico library included David Correia's *Properties of Violence*, Angela Garcia's *The Pastoral Clinic* (recommended to me by Grace Zhou), Maria Montoya's *Translating Property*, and William Kittredge's *Owning It All*.

To learn about Zuni Salt Lake, I went to the manuscripts division at the University of Utah Marriott Library. The many boxes of the Richard Hart Papers, Zuni Land Claims Papers, and the Institute of the North American West Records hold comprehensive documentation of the process of restoring Zuni Salt Lake to the tribe, including research conducted by historian Richard Hart to support the Zuni case, copies of primary source documents, committee hearing transcripts, and other relevant documents. C. Gregory Crampton's photographs from visits to Zuni country during 1966–70, also held at the Marriott Library, helped me visualize the time and place. Hart also edited a book titled *Zuni and the Courts*.

Useful secondary sources included N. H. Darton's "The Zuni Salt Lake" and John Bradbury's University of New Mexico dissertation "Origin, Paleolimnology, and Limnology of Zuni Salt Lake Maar, West-Central New Mexico." New Mexico State Museum Curator of Paleontology Spencer Lucas spoke with me about the lake's unusual geology over the phone.

Billy-Ray Belcourt's quote comes from *A History of My Brief Body*, and Zoe Todd's from the article "Fish, Kin and Hope: Tending to Water Violations in *amiskwaciwâskahikan* and Treaty Six Territory."

Chapter 6: A River Passes By Here

An essay that became the basis for this chapter was published in the *Kenyon Review Online* and as an ebook by Random House as runner-up for the 2020 Bodley Head/*Financial Times* essay prize.

The two works that inspired this chapter were Alfonso Reyes's classic essay "Visión de Anáhuac," which I read in the collection *México* (volume 1 of the Capilla Alfonsina edition of Reyes's work)—given to me by my friend Nicolás Medina Mora—and Matthew Vitz's excellent book *City on a Lake*. During the research process, Barbara Mundy's book *The Death of Aztec Tenochtitlan, the Life of Mexico City* became my most important source for pre-Hispanic and sixteenth- and seventeenth-century Tenochtitlán/Mexico City. I think of this chapter as kin to Diego Rodríguez Landeros's book *Drenajes*, the book that helped me visualize the paradox of the Mexico Basin simultaneously having excess water and water shortages.

Mundy follows Nahuatl transliteration conventions, omitting diacritic marks on the place names Tenochtitlan, Pantitlan, and Cuautitlan. I have used the Spanish transliterations common in contemporary usage. I've written the name of King Nezahualcoyotl and his dike according to contemporary convention, but left Reyes's "Netzahualcóyotl" in my quotation of his essay.

Patricia Romero Lankao's article "Water in Mexico City: What will climate change bring to its history of water-related hazards and vulnerabilities?" was beneficial for understanding the geology of the Mexico Basin. Omar Rodríguez Camarena's "Transformation and Persistence of the Basin-Valley of Mexico in the 16th and 17th Centuries" and A. K. Hernández-Espinosa's "El Sistema de Drenaje de la Ciudad de México" were helpful for pre-Hispanic and sixteenth- and seventeenth-century Mexico City. In quoting the *Cartas de Relación* of Hernán Cortés (in J. Bayard Morris's translation), I only scratched the surface of the many fascinating written accounts of the era.

For the twentieth century and contemporary water issues, I drew on numerous articles, among them "A Comprehensive Approach for the Assessment of Shared Aquifers: The Case of Mexico City" by Sandra Martínez et al., Cecilia Tortajada and Enrique Castelán's "Water Management for a Megacity: The Mexico City Metropolitan Area," Gian Carlo Delgado-Ramos's "Water and the Political Ecology of Urban Metabolism: The Case of Mexico City," Enrique Moreno Sánchez's "Lo Ambiental del Nuevo Aeropuerto Internacional de la Ciudad de México, en Texcoco, Estado de México," Dean Chahim's "Governing Beyond Capacity: Engineering, Banality, and the Calibration of Disaster in Mexico City," Francisco Platas López's "El Sistema de Drenaje Profundo de la Ciudad de México: Algunas consequencias ambientales de su diseño," Luciano Concheiro San Vicente and Xavier Nueno Guitart's "Reclaiming the (Hinter)land: Lake Texcoco and the Airport that Never Was," Alejandro

   NOTE ON SOURCES

de Coss-Corzo's "Working with the End of Water: Infrastructure, Labor, and Everyday Futures of Socio-Environmental Collapse in Mexico City," and "Historical Political Ecology in the Former Lake Texcoco: Hydrological Regulation" by Carolina Montero-Rosado et al. On the new Lake Texcoco Ecological Park, Matthew Ponsford's articles "Restoring an Ancient Lake From the Rubble of an Unfinished Airport in Mexico City," and "A Vast Wetland Park Seeks to Slake a Thirsty Megacity" provide a good overview.

I spent much of summer 2024 in the archive of CONAGUA, Mexico's national water commission, in downtown Mexico City. There, the collections I viewed included requests from local campesinos for permission to use the rivers that fed Lake Texcoco for irrigation, documents related to the construction of the Texquiquiac Tunnel and other drainage projects, to the creation of the federal zone at Lake Texcoco and the agricultural experimentation that later took place on it, to the sale of dry lakebed lands, and to Nabor Carrillo's dust management plan.

Derek Jarman's *Modern Nature* can be read slowly, like a book of hours. Matthew Gandy's article about Abney Park Cemetery, "Queer Ecology: Nature, Sexuality, and Heterotopic Alliances," was the work that convinced me of the value of queer ecology as a lens. Catriona Mortimer-Sandilands's article in *Invisible Culture*, "Unnatural Passions?: Notes Toward a Queer Ecology," provides a theorization of how the AIDS crisis influenced the development of what Mortimer-Sandilands calls a "queer ecological sensibility."

The Frank Bidart poem "Advice to the Players" is from the collection *Star Dust*; I encountered it in *Half-Light: Collected Poems 1965–2016*. In its spirit, also wrapped up in this essay are all the years I've spent biking, walking, and driving to get to know Mexico City and its water infrastructure accompanied by Mariana GJP and other friends. I'm grateful to all of them for their vicarious enthusiasm for—or at least their patience with—my obsessions.

Chapter 7: The Ecosystem that Lawsuits Made

I recall learning about the Owens Valley Water Wars in fall 2013, in the issue of *Boom: A Journal of California* that marked the centennial of the arrival of water to the Southland via the Los Angeles Aqueduct. I had previously seen images of the Owens Valley on the website of the Center for Land Use Interpretation. After reading *Boom*'s coverage, I watched the film *Chinatown*, a famous dramatization of the conflict. Shortly thereafter, after graduating from college and moving temporarily to Los Angeles, I found a copy of Mary Austin's novel *The Ford*, a speculative, feminist retelling of the events, and read it after visiting the Owens Valley. My interest was cemented when I read *The West Without Water*, by B. Lynn Ingram and Frances Malamud-Roam, and Austin's *The Land of Little Rain*.

William Kahrl's two-part article "The Politics of California Water," published in *California Historical Quarterly*, and Alexander Robinson's beautifully illustrated book *The Spoils of Dust: Reinventing the Lake that Made Los Angeles* both enriched my understanding of Owens Lake. Jay Owens's *Dust: The World in a Trillion Particles*, which came out while I was writing this book, includes two excellent chapters about the lake.

Interviews with Teri Red Owl of the Owens Valley Indian Water Commission and Phill Kiddoo—who generously spent a day guiding me around the lake—shaped this chapter. Some of the articles that helped me include Peter Vorster's "The Development and Decline of Agriculture in the Owens Valley," Jane Braxton's "Keeping the Dust Down in California's Owens Valley," and Dave Herbst and Michael Prather's "Owens Lake—From Dustbowl to Mosaic of Salt Water Habitats," published in *Lakeline*. Ro Skelton's essay "Little Starts" came out in the midst of Mariana's and my immigration process and was a salve. The Garth Greenwell quotation comes from the essay "Sex, Love, and Art in the Suburbs."

Chapter 8: Willing Sellers and the Public Trust

Interviews with Walker River Paiute Tribe Director of Land and Water Projects John McMasters, Walker River Paiute Tribe member and Mineral County extension agent Staci Emm, farmer Peter Masini, Glenn and Marlene Bunch, Peter Stanton and other members of the Walker Basin Conservancy Team, water law scholars Michael Blumm, Bret Birdsong, and Brigham Daniels, and hydrologist Jim Thomas were all crucial to this chapter. Michelle Nijhuis's article "Troubled Oasis" helped clarify the early history of the working group.

Dave Herbst's invertebrate research—with Scott W. Roberts, R. Bruce Medhurst, Ian D. Bell, Graham Chisholm, and Robert Jellison—includes the articles "Defining Salinity Limits on the Survival and Growth of Benthic Insects for the Conservation Management of Saline Walker Lake, Nevada, USA" and "Substratum Association and Depth Distribution of Benthic Invertebrates in Saline Walker Lake, Nevada, USA," along with "Walker Lake—Terminal Lake at the Brink."

The article "Using Historical General Land Office Survey Notes to Quantify the Effects of Irrigated Agriculture on Land Cover Change in an Arid Lands Watershed," by Thomas E. Dilts et al., helped me visualize the Walker Basin landscape of the past compared to that of today. The January 2011 decision by the Ninth Circuit Court of Appeals in *Walker Lake Working Group v. Walker River Irrigation District* helped me understand the trajectory and arguments of the case. I wouldn't have been able to assemble its chronology without Daniel Rothberg's excellent reporting on the case for the *Nevada Independent*.

Simeon Herskovits was ill while I was reporting this chapter and passed away shortly before the book went to press. It's a regret of mine that I was never able to speak with him about his work, which inspired so many people in the Great Basin. His colleague Iris Thornton continues the work at Walker Lake and I trust that others will carry on his legacy throughout the region.

Chapter 9: An Act of Attention

A version of this chapter first appeared as a feature story in *High Country News*, where I had the pleasure of being edited by Michelle Nijhuis. It also draws on an article I wrote for the *Guardian*'s Biodiversity section about Lake Abert, edited by Tess McClure. The chapter was also made possible by my residency at PLAYA/Summer Lake, the Waterston Desert Writing Prize, and the Oregon State University Public Humanities Collaboratory Watershed Fellowship.

I spoke about Lake Abert with Theo Dreher, Ron Larson, Todd Jarvis, Ryan Houston, and members of the Partnership for Lake Abert and the Chewaucan, including Wilson Wewa, Colleen Withers, Tess Baker, Bobby Cochran, and Jenna Stillman. On Mar Chiquita, phalaropes, and the Endangered Species Act petition, conversations with Ryan Carle, Marcela Castellino, Marina Castellino, Matías Carpinetto, Margaret Rubega, John Neil, John Luft, Patrick Donnelly, Ben Abbott, Sydney Miller, Ben Haase, Laura Josens, and Jaimi Butler all made this chapter possible.

On the critical relationship between saline lakes and the birds that depend on them, the articles I read included Sheila A. Mahoney and Joseph Jehl's "Adaptations of Migratory Shorebirds to Highly Saline and Alkaline Lakes: Wilson's Phalarope and American Avocet," Jehl's "Biology of the Eared Grebe and Wilson's Phalarope in the Nonbreeding Season: A Study of Adaptations to Saline Lakes," Maureen G. Frank and Michael R. Conover's "Threatened

Habitat at Great Salt Lake: Importance of Shallow-Water and Brackish Habitats to Wilson's and Red-Necked Phalaropes," Susan Haig's "Climate-Altered Wetlands Challenge Waterbird Use and Migratory Connectivity in Arid Landscapes," and J. Patrick Donnelly (of the USGS, not the Center for Biological Diversity)'s "Climate and Human Water Use Diminish Wetland Networks Supporting Continental Waterbird Migration."

For Lake Abert, Johnnie Moore's "Recent Desiccation of Western Great Basin Saline Lakes: Lessons from Lake Abert" was a key source. Ron Larson has written the most of anyone about Abert. His articles for *Lakeline*, *Lake Wise*, and *Western North American Naturalist* were all beneficial to me. Larson also recently published a book about Abert, *A Natural History of Oregon's Lake Abert in the Northwest Great Basin Landscape*. Finally, the Oregon Lakes Association hosted a symposium about Lake Abert that can be watched at the organization's website, which Carrie Hardison shared with me.

On phalaropes and the Laguna Mar Chiquita, Ryan Carle, Marcela Castellino, and Marina Castellino's blog posts for the Western Hemisphere Shorebird Reserve Network provided useful information, along with reports published by members of the International Phalarope Working Group: "Coordinated Phalarope Surveys at Western North American Staging Sites, 2019–2021" by Ryan Carle et al., Marcela Castellino and Arne Lesterhuis's "Wilson's Phalarope Simultaneous Census 2020: Summary and Results," and Kiki Tarr and Ryan Carle's Phalaropes and Saline Lakes Program 2023 Report. Finally, the 2024 Petition to the U.S. Fish and Wildlife Service to List Wilson's Phalarope (*Phalaropus tricolor*) Under the Endangered Species Act as a Threatened Species and to Concurrently Designate Critical Habitat, filed by the Center for Biological Diversity et al., brings together much of the information I synthesize here. Brooke Larsen was my eyes and ears at the March 2024 Phalarope Phest.

I first encountered Simone Weil and her essay "Attention and Will" during a seminar with Jennifer Rapp. The Sufjan Stevens quotation comes from an interview in *Deadline*. My citations of Mormon scripture owe a debt to Jason Brown's article "Whither Mormon Environmentalism?" Terry Tempest Williams's essay "A Totemic Act" is from the book *Erosion: Essays of Undoing*, and William Kittredge's *Hole in the Sky* was just the right book to bring to central Oregon.

Coda: The Ephemeral Forever

In its original form as an essay, this coda appeared in the journal *Shenandoah*. This essay and "A River Passes By Here," the basis of chapter 6, were both projects for "Writing the Lyric Essay: When Poetry and Nonfiction Play," an online class taught by Joanna Penn Cooper in spring 2020.

On Dorothy Napangardi and interpretations of her work, I read Napangardi's entry in "Artist Pages: Major Projects by Australian Artists at the MCA" in *Volume One: MCA Collection*, Erin Manning's *Relationscapes: Movement, Art, Philosophy*, and *Honouring and Remembering the Art and Life of Dorothy Napangardi*. The quote from John Frow comes from the book *On Interpretive Conflict*. The article "'Dreamtime' and 'The Dreaming' – An Introduction," by Christine Judith Nicholls, helped me better understand the concept of Dreaming.

On ephemeral ecology, I drew on John Jessop, Gilbert R. M. Dashorst, and Fiona M. James's *Grasses of South Australia: An Illustrated Guide to the Native and Naturalised Species*, Valerie T. Eviner's chapter "Grasslands" in *Ecosystems of California*, and the Bureau of Land Management's 1985 Yuma District Resource Management Plan.

The José Esteban Muñoz quotations come from *Cruising Utopia: The Then and There of Queer Futurity*, in which he discusses Martha Siegel's 1972 book *At the Vanishing Point: A Critic Looks at Dance*. *Mariana Valencia's Bouquet* was given to me by Rachel Kauder Nalebuff.

Both Miles Griffis's essay "Back from the Dead" about Lake Manly and David Wojnarowicz and my essay "A Cartography of Loss in the Borderlands" about the Archivo Familiar del Río Colorado appeared in *High Country News*; Miles's friendship is one of the great gifts of my time there. Arnaud de Decker and Jules Emile's "'Doomed to Stay': The Dying Villages of Mexico's Lake Cuitzeo," Abel Alvarado's "A Lake in Mexico's 'Magical Town' Is Disappearing," and Jair Ortega de la Sancha's "El Cuerpo de Agua Más Importante del Centro de México se ha Secado" provide overviews of the crises of Mexico's endorheic lakes. Limnologist Javier Alcocer of the Universidad Nacional Autónoma de México has published extensively in scientific fora about the lakes. I found my copy of Cleo Woelfe Hazard's *Underflows: Queer Trans Ecologies and River Justice* at Under the Umbrella Bookstore in Salt Lake City.

Aitmatov, Chinghiz. 1988. *The Day Lasts More Than a Hundred Years*. Translated by F. J. French. Indiana University Press.

———. 2014. *Polnoe sobraniye sochinenii v vos'mi tomax*. Uluu Toolor.

Alcocer, Javier, and U. T. Hammer. 1998. "Saline Ecosystems of Mexico." *Aquatic Ecosystem Health and Management* 1: 291–315.

Alieva, N.R. et al. 2023. "Immunological Characteristics of Diseases of the Bronchopulmonary System in Infant Children in the Aral Sea Region." *Science and Innovation* 2(9): 64–69.

Alvarado, Abel. 2024. "A Lake in Mexico's 'Magical Town' Is Disappearing." *CNN*, April 17, 2024.

Anderson, M. Kat. 2013. *Tending the Wild*. University of California Press.

Arrington, Leonard. 1966. *Great Basin Kingdom: An Economic History of the Latter-day Saints, 1830–1900*. University of Nebraska Press.

———, and Dean May. 1975. "'A Different Mode of Life': Irrigation and Society in Nineteenth-Century Utah." *Agricultural History* 49(1): 3–20.

Austin, Mary Hunter. 1997 [1917]. *The Ford*. University of California Press.

———. 1903. *The Land of Little Rain*. University of New Mexico Press.

Bagemihl, Bruce. 1999. *Biological Exuberance: Animal Homosexuality and Natural Diversity*. St. Martin's Press.

Barkley, Glenn. 2012. "Dorothy Napangardi," in *Volume One: MCA Collection*. Museum of Contemporary Art Australia.

Baxter, Bonnie K. 2018. "Great Salt Lake Microbiology: A Historical Perspective." *International Microbiology* 21:79–95.

Bean, Lowell John. 1972. *Mukat's People: The Cahuilla Indians of Southern California*. University of California Press.

Belcourt, Billy-Ray. 2020. *A History of My Brief Body*. Two Dollar Radio.

Bidart, Frank. 2005. "Advice to the Players." *Half-Light: Collected Poems 1965–2016*. Farrar, Straus and Giroux.

Bingham, Sam. 1997. *The Last Ranch: A Colorado Community and the Coming Desert*. Mariner Books.

Blackhawk, Ned. 2006. *Violence over the Land: Indians and Empire in the Early American West*. Harvard University Press.

Boom editors. 2013. "Los Angeles, Owens Valley & the Aqueduct at 100." *Boom: A Journal of California* 3(3).

Bowman, Matthew. 2012. *The Mormon People: The Making of an American Faith*. Random House.

Bradbury, John. 1967. "Origin, Paleolimnology, and Limnology of Zuni Salt Lake Maar, West-Central New Mexico." PhD diss., University of New Mexico.

Brigham D. Madsen papers. MS 0671. Special Collections, J. Willard Marriott Library, University of Utah.

Brown, Jason M. 2011. "Whither Mormon Environmental Theology?" *Dialogue* 44(2): 67–86.

Bureau of Land Management. 1985. "Yuma District Resource Management Plan." US Department of the Interior.

Cantor, Alida. 2016. "The Public Trust Doctrine and Critical Legal Geographies of Water in California." *Geoforum* 72: 49–57.

Carle, Ryan D., Gabbie Burns, Kayla Caruso, et al. 2023. "Coordinated Phalarope Surveys at Western North American Staging Sites, 2019–2022." Report of the International Phalarope Working Group.

Castellino, Marcela, and Arne Lesterhuis. 2020. "Wilson's phalarope simultaneous census 2020: Summary and results." Unpublished report of Manomet Conservation Sciences.

Center for Biological Diversity et al. 2024. "Petition to the U.S. Fish and Wildlife Service to List Wilson's Phalarope (*Phalaropus tricolor*) Under the Endangered Species Act as a Threatened Species and to Concurrently Designate Critical Habitat." March 28, 2024.

C. Gregory Crampton photograph collection, 1850–1990. P0197. Special Collections and Archives, J. Willard Marriott Library, University of Utah.

Chahim, Dean. 2022. "Governing Beyond Capacity: Engineering, Banality, and the Calibration of Disaster in Mexico City." *American Ethnologist* 49(1): 20–34.

Chekhov, Anton. 2001 [1915]. *The Steppe and Other Stories*. Translated by Roland Wilks. Penguin.

Chen, Dene-Hern. 2018. "The Country That Brought a Sea Back to Life." *BBC*, July 22, 2018.

Chinatown. 1974. Directed by Roman Polanski. Paramount Pictures.

Christy, Howard A. 1978. "Open Hand and Mailed Fist: Mormon-Indian Relations in Utah 1847–52." *Utah Historical Quarterly* 46(3): 216–35.

Concheiro San Vicente, Luciano, and Xavier Nueno Guitart. 2024. "Reclaiming the (Hinter)land: Lake Texcoco and the Airport that Never Was." In *Planetary Hinterlands*, edited by P. Gupta et al. Palgrave Studies in Globalization, Culture and Society.

Correia, David. 2012. *Properties of Violence: Law and Land Grant Struggle in Northern New Mexico*. University of Georgia Press.

Cortés, Hernando. 1928. *Five Letters: 1519–1526*. Translated by J. Bayard Morris. George Routledge and Sons.

Darton, N. H. 1905. "The Zuni Salt Lake." *Journal of Geology* 13(3): 185–93.

Davis, Diana K. 2016. *The Arid Lands: History, Power, Knowledge*. MIT Press.

Davis, Mike. 2006. *City of Quartz*. Verso.

de Coss-Corzo, Alejandro. 2022. "Working with the end of water: Infrastructure, labor, and everyday futures of socio-environmental collapse in Mexico City." *Environment and Planning E* 8(1): 171–88.

de Decker, Arnaud, and Jules Emile. 2021. " 'Doomed to Stay': The Dying Villages of Mexico's Lake Cuitzeo." *Al Jazeera*, July 25, 2021.

Delgado-Ramos, Gian Carlo. 2015. "Water and the Political Ecology of Urban Metabolism: The Case of Mexico City." *Journal of Political Ecology* 22(1): 98–114.

Didion, Joan. 1968. *Slouching Towards Bethlehem*. Farrar, Straus and Giroux.

Dilts, Thomas E., et al. 2012. "Using Historical General Land Office Survey Notes to Quantify the Effects of Irrigated Agriculture on Land Cover Change in an Arid Lands Watershed." *Annals of the American Association of Geographers* 101(3): 531–48.

Donnelly, J. Patrick, Sammy L. King, Nicholas L. Silverman, et al. 2020. "Climate and Human Water Use Diminish Wetland Networks Supporting Continental Waterbird Migration." *Global Change Biology* 26: 2042–59.

Dorn, Ed, and LeRoy Lucas. 2013 [1967]. *The Shoshoneans*. University of New Mexico Press.

duBuys, William, and Joan Meyers. 2001. *Salt Dreams: Land and Water in Lowdown California*. University of New Mexico Press.

Encyclopaedia Britannica editors. "Amu Darya." *Encyclopaedia Britannica*, March 12, 2024.

———. 2025. "Aral Sea." *Encyclopaedia Britannica*, March 19, 2025.

———. 2023. "Syr Darya." *Encyclopaedia Britannica*, July 19, 2023.

Eilers, Joe, and Ron Larson. 2014. "Introducing Terminal Lakes." *LakeLine*, Fall 2014.

Ely, Christopher. 2002. *This Meager Nature*. Northern Illinois University Press.

E. Richard Hart papers. ACCN 1251. Special Collections and Archives, J. Willard Marriott Library, University of Utah.

Eviner, Valerie T. 2016. "Grasslands." In *Ecosystems of California*. University of California Press.

Farmer, Jared. 2008. *On Zion's Mount: Mormons, Indians, and the American Landscape*. Harvard University Press.

Figes, Orlando. 2002. *Natasha's Dance*. Picador.

Foltz, Richard C. 2000. "Mormon Values and the Utah Environment." *Worldviews* 4(1): 1–19.

Fondo Aguas Nacionales, Archivo Histórico del Agua, Comisión Nacional del Agua, Mexico City.

Fondo Aprovechamientos Superficiales, Archivo Histórico del Agua, Comisión Nacional del Agua, Mexico City.

Fondo Collección Fotográfica, Archivo Histórico del Agua, Comisión Nacional del Agua, Mexico City.

Fondo Consultivo Técnico, Archivo Histórico del Agua, Comisión Nacional del Agua, Mexico City.

Fondo Infraestructura Hidráulica, Archivo Histórico del Agua, Comisión Nacional del Agua, Mexico City.

Fradkin, Philip. 2008. *Wallace Stegner and the American West*. Alfred A. Knopf.

Frank, Maureen G., and Michael R. Conover. 2019. "Threatened Habitat at Great Salt Lake: Importance of Shallow-Water and Brackish Habitats to Wilson's and Red-Necked Phalaropes." *The Condor* 121(2).

Frow, John. 2019. *On Interpretive Conflict*. University of Chicago Press.

Gallery Gondwana. 2020. "Honouring and Remembering the Art and Life of Dorothy Napangardi."

Gandy, Matthew. 2012. "Queer Ecology: Nature, Sexuality, and Heterotopic Alliances." *Environment and Planning D: Society and Space* 30(4): 727–47.

Garcia, Angela. 2010. *The Pastoral Clinic: Addiction and Dispossession Along the Rio Grande*. University of California Press.

Great Salt Lake Strike Team. 2023. "Great Salt Lake Policy Assessment." Great Salt Lake Strike Team.

Greenwell, Garth. 2022. "Sex, Love, and Art in the Suburbs." *Esquire*, April 14, 2022.

Griffis, Miles. 2024. "Learning how to live and die with long COVID." *High Country News*, June 1, 2024.

Goodfriend, Wendy. 2013. "How the Eastern Coachella Valley Used Egyptomania to Sell Dates." *KQED*, November 5, 2013.

Gorvett, Zoria. 2017. "The Deadly Germ Warfare Island Abandoned by the Soviets." *BBC*, September 26, 2017.

Gosnell, Hannah. 2020. "A Half Century of Holistic Management: What does the evidence reveal?" *Agriculture and Human Values* 37: 849–67.

Haig, Susan, M., Sean P. Murphy, John H. Matthews, et al. 2019. "Climate-Altered Wetlands Challenge Waterbird Use and Migratory Connectivity in Arid Landscapes." *Scientific Reports* 9.

Handley, George B. 2016. "Mormonism: Mormon Views of Environmental Stewardship." In *The Routledge Handbook of Religion and Ecology*, edited by Willis Jenkins, Mary Evelyn Tucker, and John Grim. Routledge.

———. 2014. "The Restoration of All Things: The Case of the Provo River Delta." In *Desert Water: The Future of Utah's Natural Resources*, edited by Hal Crimmel. University of Utah Press.

———, Terry B. Ball, and Steven L. Peck, eds. 2006. *Stewardship and the Creation: LDS Perspectives on the Environment*. Brigham Young University Religious Studies Center.

Haraway, Donna. 2003. *The Companion Species Manifesto: Dogs, People and Significant Otherness*. Prickly Paradigm Press.

Hart, John. 1996. *Storm Over Mono*. University of California Press.

Hart, Richard E., ed. 2023. *Zuni and the Courts: A Struggle for Sovereign Land Rights*. University Press of Kansas.

Hazard, Cleo Woelfe. 2022. *Underflows: Queer Trans Ecologies and River Justice*. University of Washington Press.

Herbst, David B. 2013. "Defining Salinity Limits on the Survival and Growth of Benthic Insects for the Conservation Management of Saline Walker Lake, Nevada, USA." *Journal of Insect Conservation* 17: 877–83.

———. 2012. "Substratum Association and Depth Distribution of Benthic Invertebrates in Saline Walker Lake, Nevada, USA." *Hydrobiologia* 700: 61–72.

———, R. Bruce Medhurst, Ian D. Bell, et al. 2014. "Walker Lake—Terminal Lake at the Brink." *LakeLine*, Fall 2014.

———, and Michael Prather. 2014. "Owens Lake—From Dustbowl to Mosaic of Salt Water Habitats." *LakeLine*, Fall 2014.

Hernández-Espinosa, A. K., E. M. Otazo-Sánchez, et al. 2021. "El Sistema de Drenaje de la Ciudad de México." *Publicación Semestral Pädi* 9(17): 24–30.

Hess, Ralph. 1912. "The Beginnings of Irrigation in the United States." *Journal of Political Economy* 20(8): 807–33.

Hopkins, Gerard Manley. 1953 [1877]. "Pied Beauty." In *Gerald Manley Hopkins: Poems and Prose*. Penguin Classics.

Houghton, Samuel. 1976. *A Trace of Desert Waters: The Great Basin Story.* Arthur H. Clark Company.

Ingram, B. Lynn, and Frances Malamud-Roam. 2013. *West without Water: What Past Floods, Droughts, and Other Climatic Clues Tell Us about Tomorrow.* University of California Press.

Institute of the North American West records, 1870–1993. ACCN 1293. Special Collections and Archives, J. Willard Marriott Library, University of Utah.

James E. Talmage papers. MSS 229. L. Tom Perry Special Collections, Harold B. Lee Library, Brigham Young University.

Jarman, Derek. 1991. *Modern Nature.* Vintage.

Jehl, Joseph R., Jr. 1985. "Biology of the Eared Grebe and Wilson's Phalarope in the Nonbreeding Season: A Study of Adaptations to Saline Lakes." *Studies in Avian Biology* 12.

Jellison, Robert, et al. 2008. "Salt Lakes: Values, Threats and Future." In N. V. C. Polunin, ed., *Aquatic Ecosystems*. Cambridge University Press.

Jessop, John, Gilbert R. M. Dashorst, and Fiona M. James. 2006. *Grasses of South Australia: An Illustrated Guide to the Native and Naturalised Species.* Wakefield Press.

John Steele collection, 1989–1990. UA 5572, Series 2 File 127, Carton 11, Folder 10. L. Tom Perry Special Collections, Harold B. Lee Library, Brigham Young University.

Kahrl, William. 1976. "The Politics of California Water." *California Historical Quarterly* 55(2): 98–120.

Kamp, Marianne. 2024. *Collectivization Generation: Oral Histories of a Social Revolution in Uzbekistan.* Cornell University Press.

———. 2022. "Soviet Collectivization in Central Asia." *Oxford Research Encyclopedia of Asian History*, November 22, 2022.

Kittredge, William. 1992. *Hole in the Sky*. Penguin Random House.

————. 1987. *Owning It All*. Graywolf Press.

Koch, Natalie. 2023. *Arid Empire: The Entangled Fates of Arabia and Arizona*. Verso.

KQED. 2014. "Plagues and Pleasures at the Salton Sea." September 5, 2014. YouTube video.

KQED Deep Look. 2023. "This Daring Fly Swims in Shimmering Bubble Shield." YouTube video.

Larsen, Leia. 2023. "A Utah Lake dredging critic got dragged into a lawsuit with a private company. Did he also lose a public grant?" *Salt Lake Tribune*, November 8, 2023.

Larson, Ron. 2023. "American Avocets Flock to Oregon's Eastside Lakes." *Lake Wise* newsletter, October 2023.

————. *A Natural History of Oregon's Lake Abert in the Northwest Great Basin Landscape*. University of Nevada Press.

————. 2016. "Recent Desiccation-Related Ecosystem Changes at Lake Abert, Oregon: A Terminal Alkaline Salt Lake." *Western North American Naturalist* 76(4): 389–404.

————, and Joe Eilers. 2014. "Lake Abert, OR: A Terminal Lake Under Extreme Water Stress." *LakeLine*, Fall 2014.

Lauener, Paul. 2011. "A Sea Returns to Life, a Sea Slowly Dies." *New Internationalist*, November 1, 2011.

Little, Jane Braxton. 2015. "Keeping the Dust Down in California's Owens Valley." *Los Angeles Times*, March 2, 2015.

Mahoney, Sheila A., and Joseph Jehl. 1985. "Adaptations of Migratory Shorebirds to Highly Saline and Alkaline Lakes: Wilson's Phalarope and American Avocet." *The Condor* 87(4): 520–27.

Manning, Erin. 2009. *Relationscapes: Movement, Art, Philosophy*. MIT Press.

Marks, Robert. 2023. "Before Mulholland: Land, Water, and Power in the Mono Basin, 1872–1923." *Southern California Quarterly* 105(2): 105–41.

————. 2023. "Dispossessed Again: Paiute Land Allotments in the Mono Basin 1907–1929." *Eastern Sierra History Journal* 4.

Martínez, Sandra, et al. 2015. "A Comprehensive Approach for the Assessment of Shared Aquifers: The Case of Mexico City." *Sustainable Water Resources Management* 1: 111–23.

McLaren, William. 2015. "A Fishery, a Sanctuary, a Sink, and a Disaster: The Often-Hapless Management of California's Salton Sea." *UC Law Environmental Journal* 21(2): 141–60.

McPhee, John. 1981. *Basin and Range*. Farrar, Straus and Giroux.

McWilliams, Carey. 1949. *California: The Great Exception*. University of California Press.

———. 2000. *Factories in the Field: The Story of Migratory Farm Labor in California*. University of California Press.

———. 1946. *Southern California: An Island on the Land*. Peregrine Smith.

Montoya, Maria E. 2002. *Translating Property: The Maxwell Land Grant and the Conflict Over Land in the American West, 1840–1900*. University of California Press.

Montero-Rosado, Carolina, et al. 2023. "Historical Political Ecology in the Former Lake Texcoco: Hydrological Regulation." *Land, MDPI* 12(5): 1–31.

Moore, Johnnie N. 2016. "Recent Desiccation of Western Great Basin Saline Lakes: Lessons from Lake Abert, Oregon, U.S.A." *Science of the Total Environment* 554–55: 142–54.

Moreno Sánchez, Enrique. 2017. "Lo Ambiental del Nuevo Aeropuerto Internacional de la Ciudad de México, en Texcoco, Estado de México." *Letras Verdes. Revista Latinoamericana De Estudios Socioambientale* 22: 248–73.

Mortimer-Sandilands, Catriona. 2005. "Unnatural Passions?: Notes Toward a Queer Ecology." *Invisible Culture*, issue 9.

Mundy, Barbara E. 2015. *The Death of Aztec Tenochtitlan, the Life of Mexico City*. University of Texas Press.

Muñoz, José Esteban. 2009. *Cruising Utopia: The Then and There of Queer Futurity*. New York University Press.

Myron L. and Lucille P. Sutton papers. ACCN 1763. Special Collections, J. Willard Marriott Library, University of Utah.

New Internationalist editors. 2017. "Death and Re-Birth of a Lake: How Water Came Back to the Dry Aral Sea." *New Internationalist*, June 14, 2017.

Nicholls, Christine Judith. 2014. "'Dreamtime' and 'The Dreaming' – An Introduction." *The Conversation*, January 22, 2014.

Nijhuis, Michelle. 1999. "Troubled Oasis." *High Country News*, September 13, 1999.

Obertreis, Julia. 2017. *Imperial Desert Dreams: Cotton Growing and Irrigation in Central Asia, 1860–1991*. Vandenhoeck and Ruprecht.

Oklahoma State University Library. n.d. "International Criminal Court—Research and Current Cases." Last modified April 15, 2025.

Oregon Lakes Association. "Lake Abert Colloquium." YouTube video.

Ortega de la Sancha, Jair. 2023. "El Cuerpo de Agua Más Importante del Centro de México se ha Secado." *Gatopardo*, August 11, 2023.

Owen, David. 2017. *Where the Water Goes: Life and Death Along the Colorado River*. Riverhead Books.

Owens, Jay. 2023. *Dust: The Modern World in a Trillion Particles*. Harry N. Abrams.

Park, Benjamin. 2024. *American Zion: A New History of Mormonism*. Liveright.

Parry, Darren. 2019. *The Bear River Massacre: A Shoshone History*. By Common Consent Press.

Pasternak, Boris. 1997 [1957]. *Doctor Zhivago*. Translated by Max Hayward and Manya Harari. Pantheon.

Pavlik, Bruce M. 2008. *The California Deserts: An Ecological Rediscovery*. University of California Press.

Peterson, Maya K. 2019. *Pipe Dreams: Water and Empire in Central Asia's Aral Sea Basin*. Cambridge University Press.

Platas López, Francisco. 2008. "El sistema de drenaje profundo de la Ciudad de México. Algunas consecuencias ambientales en su diseño." *Investigación y Diseño: Anuario del Posgrado 05*. Universidad Autónoma Metropolitana–Unidad Xochimilco.

Platonov, Andrei. 2008 [1999]. *Soul*. Translated by Robert and Elizabeth Chandler with Katia Grigoruk, Angela Livingstone, Olga Meerson, and Eric Naiman. New York Review Books.

Ponsford, Matthew. 2023. "Restoring an Ancient Lake from the Rubble of an Unfinished Airport in Mexico City." *MIT Technology Review*, February 13, 2023.

———. 2024. "A Vast Wetland Park Seeks to Slake a Thirsty Megacity." *Bloomberg CityLab*, August 4, 2024.

Qobil, Rustam. 2015. "Waiting for the Sea." *BBC*, February 25, 2015.

Reisner, Marc. 1986. *Cadillac Desert: The American West and Its Disappearing Water*. Penguin.

Reyes, Alfonso. 2005. "Visión de Anáhuac." In *México*. Fondo de Cultura Económica.

Robinson, Alexander. 2018. *The Spoils of Dust: Reinventing the Lake that Made Los Angeles*. Applied Research & Design.

Rodríguez, Richard. 1993. *Days of Obligation: An Argument with My Mexican Father*. Penguin.

———. 1983. *The Hunger of Memory*. Bantam.

Rodríguez Camarena, Omar. 2022. "Transformation and Persistence of the Basin-Valley of Mexico in the 16th and 17th Centuries." *Journal of Interdisciplinary History of Ideas* 11(22).

Rodríguez Landeros, Diego. 2022. *Drenajes.* Almadía.

Romero Lankao, Patricia. 2010. "Water in Mexico City: What will climate change bring to its history of water-related hazards and vulnerabilities?" *Environment & Urbanization* 22(1): 157–78.

Rothberg, Daniel. 2018. "9th Circuit ruling on Walker Lake puts far-reaching water rights issue before Nevada Supreme Court." *Nevada Independent,* May 27, 2018.

———. 2020. "The Nevada Supreme Court Ruled Against Reshuffling Water Rights to Fix Environmental Issues. Walker Lake Advocates Still See a Path Forward." *Nevada Independent,* September 4, 2020.

———. 2021. "9th Circuit to District Court: Consider 'Public Trust' Remedies in Walker Lake Case." *Nevada Independent,* February 4, 2021.

Salton Sea Environmental Timeseries. n.d. saltonseascience.org.

Saltykov-Shchedrin, Mikhail. 1988 [1880]. *The Golovlyov Family.* Translated by Ronald Wilks. Penguin Classics.

Savory, Allan. 1988. *Holistic Resource Management.* Island Press.

Sax, Joseph. 1970. "The Public Trust Doctrine in Natural Resource Law: Effective Judicial Intervention." *Michigan Law Review* 68(3).

Sayre, Nathan F. 2001. *The New Ranch Handbook: A Guide for Restoring Western Rangelands.* Quivira Coalition.

———. 2017. *The Politics of Scale: A History of Rangeland Science.* University of Chicago Press.

Schorr, David. 2012. *The Colorado Doctrine: Water Rights, Corporations, and Distributive Justice on the American Frontier.* Yale University Press.

Scott, A. O. 2012. "Taking Flight from Misery, Only to Court Peril." *New York Times,* August 28, 2012.

Shamshetova, Anjim. 2023. "Ecological Problems of the Aral Sea Region and the Psychology of the Population." *Science and Innovation* 2(4): 515–22.

Siegel, Martha. 1972. *At the Vanishing Point: A Critic Looks at Dance.* Dutton.

Skelton, Rose. 2021. "Little Starts." *Ecotone* 29.

Starrs, Paul F. 1988. *Let the Cowboy Ride: Cattle Ranching in the American West.* Johns Hopkins University Press.

Steele, Lauren. 2021. "The Water That Couldn't Save." *Deseret News,* April 2021.

Stegner, Wallace. 1987. *The American West as Living Space.* University of Michigan Press.

———. 1998. "At Home in the Fields of the Lord." In *Marking the Sparrow's*

Fall: Wallace Stegner's American West, edited by Page Stegner. Henry Holt & Co.

Steward, Julian H. 1938 [1997]. "Basin-Plateau Aboriginal Sociopolitical Groups." University of Utah Press.

Tarr, David, and Eskender Trushin. 2004. "Did the Desire for Cotton Self-Sufficiency Lead to the Aral Sea Disaster?" World Bank.

Tarr, Kiki, and Ryan Carle. 2023. "Phalaropes and Saline Lakes Program 2023 Report." Unpublished report by Oikonos Ecosystem Knowledge.

Teichmann, Christian. 2008. "Canals, Cotton, and the Limits of De-Colonization in Soviet Uzbekistan, 1924–41." *Central Asian Survey* 26(4): 499–519.

Thomas, Alexander. 2002. "Irrigating the Mormon Heartland: The Operation of the Irrigation Companies in Wasatch Oasis Communities, 1847–1880." *Agricultural History* 76(2): 172–87.

Thomas Caldwell Adams papers. MS 0043. Special Collections, J. Willard Marriott Library, University of Utah.

Thomas Day Jr. Collection. MSS 2071. L. Tom Perry Special Collections, Harold B. Lee Library, Brigham Young University.

Todd, Zoe. 2017. "Fish, Kin and Hope: Tending to Water Violations in amiskwaciwâskahikan and Treaty Six Territory." *Afterall*, March 7, 2017.

Tolstoy, Leo. 2002 [1878]. *Anna Karenina*. Translated by Richard Pevear and Larissa Volokhonsky. Penguin.

———. 2008 [1869]. *War and Peace*. Translated by Richard Pevear and Larissa Volokhonsky. Vintage Classics.

Tortajada, Cecilia, and Enrique Castelán. 2003. "Water Management for a Megacity: The Mexico City Metropolitan Area." *Ambio* 32(2): 124–29.

Tracey, Caroline. 2024. "A Cartography of Loss in the Borderlands." *High Country News*, February 21, 2024.

———. 2023. "How the tiny brine shrimp can help protect the Great Salt Lake." *High Country News*, May 18, 2023.

———. 2023. "In Search of Answers at the Salton Sea." *High Country News*, June 1, 2023.

———. 2023. "LDS Environmentalists Want Their Institution to Address the Great Salt Lake's Collapse." *High Country News*, January 24, 2023.

———. 2022. "Utah's youth climate activists held a funeral for the Great Salt Lake." *High Country News*, September 16, 2022.

Utah Physicians for a Healthy Environment. 2024. "Great Salt Lake Rally at the Utah State Capitol January 20, 2024." YouTube video.

Utichi, Joe. 2017. "Sufjan Stevens Nearly Played the Narrator in Luca Guadag-
    nino's 'Call Me By Your Name.'" *Deadline*, December 13, 2017.

Valencia, Mariana. 2019. *Bouquet*. 3 Hole Press.

Vasyakina, Oksana. 2023. *Wound*. Translated by Eline Alter. Catapult.

Virgil V. Peterson papers. MSS 1945. L. Tom Perry Special Collections, Harold
    B. Lee Library, Brigham Young University.

Vitz, Matthew. 2018. *A City on a Lake: Urban Political Ecology and the Growth
    of Mexico City*. Duke University Press.

Vollmann, William. 2009. *Imperial*. Penguin.

Vorster, Peter. 1992. "The Development and Decline of Agriculture in the
    Owens Valley." In *The History of Water: Eastern Sierra Nevada, Owens Val-
    ley, White–Inyo Mountains*, edited by Clarence A. Hall Jr., Victoria Doyle-
    Jones, and Barbara Widawski. University of California, White Mountain
    Research Station.

Voyles, Traci Brynne. 2022. *The Settler Sea: California's Salton Sea and the Con-
    sequences of Colonialism*. University of Nebraska Press.

Wang, Jida, Chunqiao Song, et al. 2018. "Recent global decline in endorheic
    basin water storages." *Nature Geoscience* 11: 926–32.

Walker Lake Working Group v. Walker River Irrigation District. No. 15–16342
    (2021).

Wallace F. Bennett Papers. 1966. MSS 20. L. Tom Perry Special Collections,
    Harold B. Lee Library, Brigham Young University.

Waters, Frank. 1985. *The Colorado*. Swallow Press.

Weil, Simone. 1952 [1947]. "Attention and Will." In *Gravity and Grace*. Trans-
    lated by Emma Craufurd. Routledge.

Wheeler, William. 2021. *Environment and Post-Soviet Transformation in Kazakh-
    stan's Aral Sea Region*. University College London Press.

Whish-Wilson, Philip. 2002. "The Aral Sea Environmental Health Crisis." *Jour-
    nal of Rural and Tropical Public Health* 1(2): 29–34.

Willard and Celia Luce Collection. MSS 6784. L. Tom Perry Special Collec-
    tions, Harold B. Lee Library, Brigham Young University.

Williams, Raymond. 1973. *The Country and the City: The Greening of the San
    Francisco Bay Area*. University of Washington Press.

Williams, Terry Tempest. 2023. "I Am Haunted by What I Have Seen at the
    Great Salt Lake." *New York Times*, March 25, 2023.

———. 1991. *Refuge: An Unnatural History of Family and Place*. Pantheon.

———. 2019. "A Totemic Act." In *Erosion: Essays of Undoing*. Macmillan.

Williams, William D. 2002. "Environmental Threats to Salt Lakes and the Likely Status of Inland Saline Ecosystems in 2025." *Environmental Conservation* 29(2).

Wilson, Janet, and Nat Lash. 2023. "The 20 Farming Families Who Use More Water from the Colorado River Than Some Western States." *ProPublica*, November 9, 2023.

Wine, Michael L. 2022. "Irrigation water use driving desiccation of Earth's endorheic lakes and seas." *Australasian Journal of Water Resources*, 28(1): 74–85.

Winkler, David W. 1977. "An Ecological Study of Mono Lake, California." Institute of Ecology, University of California–Davis.

Worster, Donald. 1985. *Rivers of Empire*. Oxford University Press.

Wurtsbaugh, Wayne, and Maciej Gliwicz. 2001. "Limnological Control of Brine Shrimp Population Dynamics and Cyst Production in the Great Salt Lake, Utah." *Hydrobiologia* 466: 119–32.

Wurtsbaugh, Wayne A., et al. 2017. "Perspective: Decline of the World's Saline Lakes." *Nature Geoscience* 10: 816–21.

Yin, Steph. "A Warm and Wet Spring, Then 200,000 Dead Saigas." *New York Times*, January 17, 2018.

Yogi, Stan, ed. 1996. *Highway 99: A Literary Journey Through California's Great Central Valley*. Heyday Books.

Zimmer, Carl. 2015. "Death on the Steppes: Mystery Disease Kills Saigas." *New York Times*, May 29, 2015.

———. 2015. "More Than Half of Entire Species of Saigas Gone in Mysterious Die-off." *New York Times*, November 2, 2015.

Zuni Land Claims papers. ACCN 1200. Special Collections and Archives, J. Willard Marriott Library, University of Utah.